The Science Behind a Happy Dog

To Chris, Miriam & Otto —
I hope you enjoy this
and find it useful —
Very best wishes to you all,

The Science Behind a Happy Dog

Canine Training, Thinking and Behaviour

Emma K. Grigg, Tammy M. Donaldson

Illustrations by Rory Walker and H. Light

First published 2017

Copyright © Emma K. Grigg, Tammy M. Donaldson 2017

Illustrations by Rory Walker

All rights reserved. No part of this publication may be reproduced, stored in a retrieval system, or transmitted, in any form or by any means, electronic, mechanical, photocopying, recording or otherwise, without prior permission of the copyright holder.

Published by
5M Publishing Ltd,
Benchmark House,
8 Smithy Wood Drive,
Sheffield, S35 1QN, UK
Tel: +44 (0) 1234 81 81 80
www.5mpublishing.com

A Catalogue record for this book is available from the British Library

ISBN 978-1-910455-75-3

Book layout by Servis Filmsetting Ltd, Stockport, Cheshire
Printed by Replika Press Pvt Ltd, India
Photos by Emma Grigg, Tammy Donaldson unless otherwise indicated

Contents

Dedication		vi
About the authors		vii
Chapter 1	Introduction	1
Chapter 2	Quality of life in dogs	13
Chapter 3	Canine cognition: how your dog sees the world	31
Chapter 4	Reading your dog	51
Chapter 5	The science of humane dog training	71
Chapter 6	Canine mental health: the importance of enrichment	105
Chapter 7	Canine behavioral problems and solutions	127
Chapter 8	Canine physical wellness	167
Chapter 9	Veterinary care for your happy dog	195
Chapter 10	Too short lives: quality of life and end-of-life decisions for your dog	225
Annotated Bibliography: a brief list of some of our favorite dog resources		249
Index		257

Dedications

To my parents, who shared with me their love and respect for all creatures, great and small; to my family, for their support throughout it all; and to all the lovely dogs who have added so much to my life over the years. In particular, this book is in memory of our Miss Maudie, who changed it all. I still miss you, my happy girl. – E.K.G.

This book would not be possible without the profound love and inspiration of my husband, Nathan and my daughter, Gabriella who helped me realize the splendid significance of happiness. I dedicate this book to the memory of the two best teachers a person could have, Mojo and Aloysius. These two dogs sometimes frustrated me and constantly challenged me to further my knowledge of everything canine in order to create the happy lives they deserved and in doing so nurtured a love for them and for dogs that will forever endure. – T.M.D.

About the authors

Emma K. Grigg, PhD, is a Certified Applied Animal Behaviorist, an adjunct professor of canine behavior at Bergin University of Canine Studies (Rohnert Park, CA), and a post-doctoral Research Associate at the University of California, Davis. Previously, she taught companion animal behaviour at Ross University School of Veterinary Medicine, and has authored a number of scientific publications on canine, feline, and marine mammal behaviour. She currently lives in northern California with her husband, son, four cats, and one slightly neurotic "island dog."

Photo: © Sherri Rieck

Tammy McCormick Donaldson, PhD, has been a Certified Applied Animal Behaviorist since 2003, specializing in aggression in dogs. Her PhD research focused on behavioral evaluation for the detection of aggression in dogs. Her passion is welfare of all animals in rescue and she currently serves as president of The Horse Protection League (Arvada, CO). She lives with her husband, daughter, two happy dogs and a lovely horse in Golden, CO.

Courtesy of Treva Copeland

Chapter 1

Introduction

Photo: iStock

I can't imagine a world without dogs. I have shared most of my life with dogs; at times when this wasn't possible, my home always seemed strangely empty. Dogs have provided me with comfort, companionship, nearly endless entertainment, exercise, protection, and much joy (as well as, occasionally, significant frustration, and great sadness at their loss). Growing up, our family always had dogs (and cats) in our home; still, I was so captivated by dogs that as a young child, I would point each and every one out to my mother, wherever we went (and whenever we went there). This went on for many years. One day after I had joyfully pointed out yet another passing dog, she gave a small sigh and looked at me with amused indulgence (or was it mild exasperation?), saying, "Yes, dear, there are a *lot* of dogs in the world."

Figure 1.1: My older sister and the first family dog in my life, Hector, a lovely, gentle Old English Sheepdog. (Photo: C. Grigg)

Figure 1.2: One of my canine soulmates, Elliott, in a quiet moment. (Photo: E. Grigg)

Dogs in our lives

She was right; there are a *lot* of dogs in the world. In a 2012 analysis, the American Veterinary Medical Association reported that in the US alone, there were approximately 70 million dogs, and over 43 million households (36.5 percent of all US households) included at least one dog.[1] There are likely even higher numbers now. In 2015 in the UK, there were an estimated 8.5 million companion dogs, with 24 percent of households including at least one dog.[2] And, humans have shared their lives with dogs, to varying degrees, for a very long time. Recent research indicates that dogs were likely domesticated between 14,000 and 40,000 years ago, perhaps longer, and descended from Eurasian wolves. The closest living relatives of modern domestic dogs are the grey wolves, although some scientists suggest that the ancestor of today's dogs was a now-extinct cousin of the present-day grey wolf. How, where,[3] and when dogs were first domesticated are surprisingly touchy subjects among many dog scientists and geneticists; a recent feature in *Science* magazine referred to these discussions as the "dog wars" (Grimm 2015). Setting aside the hotly debated question of where and when the process occurred, the first "domestic" dogs were most likely scavengers on human populations;[4] not a surprise to anyone who has lost a

---

1 American Veterinary Medical Association, *US Pet Ownership and Demographics Sourcebook*, 2012.
2 The Pet Food Manufacturers' Association *Pet Population Report*, available at www.pfma.org.uk/pet-population-2015.
3 Recent research concludes that the modern domestic dog likely first appeared in Central Asia; see the Shannon et al. 2015 reference at the end of the chapter. However, it is quite possible that dogs were domesticated more than once in the course of human history, in a number of different parts of the world. Perhaps not all of these "domestication events" were successful, but one or more of them eventually resulted in our modern day domestic dog.
4 An alternate but less widely accepted theory suggests that domestication of dogs occurred due to both parties' realization of the mutual benefits of cooperative hunting, with the ancestral dogs being better at locating and immobilizing prey, followed by the human hunters better equipped to kill large prey, eventually leading to food sharing by humans with their canine team members.

sandwich or piece of pizza through a moment's inattention in the presence of a hungry dog. Currently, the leading theories of domestication suggest that, in the dogs' ancestors, tolerance for humans (including reduced fear of and aggression towards humans) was an important trait, and selection for this trait was a major factor shaping today's domestic dog. The most tolerant dogs were able to live in closest proximity to, and get the most food from, bands of human hunter–gatherers; and thus were able to reproduce (and pass on their genes) more successfully. As time went on, dogs evolved a suite of social behaviors which enabled them to interact more successfully with humans, including being able to quickly bond with humans (in less than 20 minutes in some studies) and to accurately interpret human social cues (including pointing gestures, and the human gaze). Modern dogs display, as the canine researcher Adam Miklósi puts it, a sophisticated social competence, defined as "the ability of individuals to harmonize their needs and expectations with those of the collective, letting them fit in with the group" (Miklósi 2015, p. 17). This ability is even more impressive when we consider that dogs are able to exhibit social competence with a species completely different from their own, i.e., humans! We will talk more about canine cognitive abilities in Chapter 3.

At this point, however, it may be most important to remember that, while modern domestic dogs are certainly related to wolves and share some similar traits, they are most definitely not wolves. The process of domestication had very significant and important effects on how domestic dogs behave, and how they perceive and interact with humans; understanding the modern domestic dog requires consideration of these effects. In a long-running study of selective breeding for "tameness" in Russian silver foxes, researchers discovered that friendliness towards humans was indeed heritable, and that many physical traits not seen in wild foxes (but very similar to those seen in domestic dogs) came along with tame behavior (without deliberate selection for the physical traits). For more on this very interesting, often-cited study, see the Trut 1999 reference at the end of this chapter. In addition, our dogs know that we are not dogs; the research overwhelmingly supports that, while dogs have maintained the ability to socialize with other dogs while also socializing with humans, they see the two interactions differently. For this reason, social dominance (while a real phenomenon within many dog groups) does not strictly apply to human–dog relationships.[5] Rather, dogs appear to view us as social partners, who often provide important information about where and how to acquire food (an essential resource), and who also can be a source of emotional support and attachment. We'll talk a lot more about this, and the practical importance of this in living with and training our dogs, in Chapters 3 and 5.

The many benefits of a life shared with dogs

These days, dogs are our companions, they work with us and for us, they look to us for care and support, and they help keep us safe from many of life's hazards. Specially-trained working dogs serve a variety of functions in human society, ranging from (to name just a few roles) livestock handling and care; to military and law-enforcement dogs who detect bombs, fugitives,

---

[5] For a nice review of our current understanding of the human–dog bond, see Payne et al. 2015 or Beck 2014, referenced at the end of this chapter.

or drugs; to search and rescue dogs who save people trapped by natural disasters such as earthquakes or avalanches. Guide and service dogs provide invaluable assistance to visually- and hearing-impaired members of our society, as well as people with limited mobility, and in recent years are being trained to help autistic children interact with their worlds. In addition to all of this, we are learning that they can help keep us healthy, both physically and mentally, simply by being dogs. At the time of writing this book, well-documented health benefits of pet ownership (and effects are often greatest for living with dogs) include improvements in cardiovascular health (lower systolic blood pressure and blood cholesterol levels, better survival rates after heart attack); and in mental health (lower levels of mental stress, fewer feelings of loneliness and depression, and higher self-esteem). On a broader scale, studies of people interacting with dogs (known officially as human–animal interaction, or HAI) have documented positive health effects in the areas of social behavior, interpersonal interactions, and mood; improvement in scores (reflecting lower stress) in stress measures such as cortisol levels (a prime physiological indicator of stress in both dogs and humans), heart rate, and blood pressure; and reductions in feelings of fear and anxiety. Although less well-documented, at least at the time of publication, other benefits of HAI appear to include reduction of levels of stress-related neurotransmitters such as epinephrine and norepinephrine, improved immune system functioning and pain management; increased trustworthiness of and trust towards other persons; reduced aggression; and even enhanced empathy and improved learning (see Beetz et al. 2012a, for a complete review of these studies). Dr. Erika Friedmann, who has published extensively on the health benefits of pet companionship for humans, notes that a person's physical health is strongly influenced by his/her social and psychological health, and suggests that one main benefit of pets may be mediating the long-term effects of stress (i.e., by moderating or serving as buffers in the relationship between life's often-unavoidable stressors and our physical and emotional health; Friedmann and Son 2009). In 2012, researchers reported that children with attachment disorders facing a stressful situation (such as giving a presentation to a group of adults) experienced significantly lower stress when they faced the situation with a therapy dog at their side (even lower than when they had a friendly human or stuffed toy with them; Beetz et al. 2012b). A similar phenomenon was noticed many years earlier by Dr. Boris Levinson, a psychologist who worked with emotionally withdrawn children. In 1961 he reported to the American Psychological Association that dogs were able to help him reach these children, even when previous human therapy attempts had failed; with his own dog present, Dr. Levinson was able to help these children much more effectively. Dr. Levinson's work is described in much greater detail in his book, *Pet-Oriented Child Psychotherapy*, referenced at the end of this chapter. Dr. Levinson is often referred to as "the father of animal-assisted therapy". Pets provide social support, defined as the physical and emotional support given to us by our family, friends, coworkers, and others (Beck 2014), much like that provided by human companions. For many people, pets can provide greater (and more consistent) social support than other humans. Simply put, we feel less lonely when a dog is by our side.

Some researchers have suggested that these benefits go beyond the individual dog owner,[6] to

---

[6] A note on wording: when writing this book, we debated about which term to use to refer to the human members of canine households. Some readers may not be comfortable with the use of the terms "pet" or "owner," and prefer terms

benefit human society as a whole, by increasing what they term "social capital": the features of our social lives (such as our networks, societal norms, and trust in our fellow man) that enable individuals to work together to achieve shared goals, and to promote cooperation for mutual benefit (Wood et al. 2005). Pet ownership has been associated with increased social contact with others, civic engagement, and better perception of the friendliness of the neighborhood. I think of this as the "dog park effect": pet ownership can provide regular opportunities for interacting in a friendly way with our neighbors, and with other like-minded dog owners. A good friend of mine was an unfamiliar neighbor when we first met; earlier in that week she had adopted the littermate of our puppy Elliott, who we had just adopted that day from the local animal shelter, and she recognized Elliott as we walked past her house. Both Elliott and his brother Chili are gone now, and we miss them still, but our human friendship remains.

> **Box 1.1: Oxytocin: the "love hormone"?**
>
> Although best known for its role in childbirth and breastfeeding, oxytocin is sometimes called "the love hormone" for its emotional effects – it plays a role in positive social bonding (between mother and child, as well as romantic attachment and empathy; and in other species as well as humans), can reduce anxiety, and even increases pain thresholds. One journalist recently described the oxytocin-fueled bond between owners and their dogs as "a feedback loop of love."* These effects may help explain the ability of our relationships with our dogs to protect or buffer us from many of the negative effects of stress in our daily lives, as well as the tendency of our dogs to look to us for support and guidance in many situations. Familiar humans (such as the dog's caretaker) are more reassuring to dogs in stressful situations than other, less familiar humans.
>
> *Jan Hoffman, April 16 2015, in The New York Times; see also Nagasawa et al. 2009.

The bond between dog and human: we both benefit

Much of the research done on the social bonds between human and dog point to the probable role of oxytocin, a hormone produced in the brains of both humans and dogs, in the formation and maintenance of the human–animal bond, such as the bond between humans and their canine companions. For more on oxytocin, see Box 1.1. Many scientists note that the partnership between domestic dogs and their humans most resembles the attachment bonds that characterize human caregiver–infant relationships. A 2014 study investigated the brain activity of human mothers as they looked at their own dog and their own child, as well as at unfamiliar dogs and children, and found striking similarities in the brain activity when mothers looked at their own child and their own dog (similarities not seen when the women looked at unfamiliar dogs or children), suggesting parallels in the relationship between humans and their children, and between humans and their dogs (Stoeckel et al. 2014). Dogs regularly display behaviors

---

that better capture the relationship between our companion animals and ourselves. We use the term owner (along with other terms such as caretaker or guardian), as this is legally accurate in most cases, but we do so in the belief that with "ownership" of any animal comes the responsibility for providing that animal with the best care possible, to the best of our ability.

Figure 1.3: "Smiling" dog gaze. (Photo: S. Watko); relaxed dog. (Photo: E. Grigg)

characteristic of human–human attachment relationships, including proximity seeking (moving to seek out a familiar figure in response to a stressful situation). As is seen with small children, the absence of a familiar person will often cause marked distress in a dog (magnified in some dogs to full-blown separation anxiety, which we'll talk about more in Chapter 6). The presence of a human can help mitigate the effects of stressful events on dogs, as has been demonstrated by studies of cortisol levels, even in dogs living in shelters (a very stressful situation for dogs); in human attachment theory, this phenomenon is known as the "safe haven" effect (Gácsi et al. 2013). The presence of a familiar human gives dogs the confidence to investigate novel objects or surroundings (a phenomenon known as the "secure base effect").[7] The human–animal bond is a two-way relationship; both humans and dogs benefit. The similarity between the dog–human bond and the parent–child bond goes hand in hand with the theory that modern dogs have been selected for juvenile features (features like large eyes, short snouts, playful behavior; think of all the traits that make puppies so irresistible to so many of us; McGreevy and Nicholas 1999). This retention of juvenile features in dogs, puppy-like features often preferred by humans, is known as neoteny.[8] Juvenile features in mammals are believed to discourage aggression and encourage compassion and nurturing behaviors; and humans do generally prefer dogs with more juvenile facial features, making it plausible that juvenile features provide dogs living with humans with a selective advantage over their fellow dogs.

---

7 For more information and a nice review of research on many aspects of the human–canine bond, see Kaminski and Marshall-Pescini's 2014 book *The Social Dog: Behavior and Cognition*.

8 See Beck 2014 for a more in-depth discussion of the role of neoteny in the domestication of dogs; the father of modern ethology, Konrad Lorenz, was one of the first to suggest (over 70 years ago) that juvenile features stimulate nurturing behavior in mammals.

The science of dogs

Now is a great time to be thinking (or reading) about dogs, from a social and scientific standpoint, as there is currently a vast and even overwhelming array of scientifically-sound information available on working and living with dogs. My co-author and I are scientists by training, and have taught about and worked with dogs (and cats) with and without behavioral issues for a combined total of almost 20 years. When we committed to writing this book, we discovered that the many wonderful (and few not so wonderful) dog books and resources out there with which we were familiar, were in fact only the tip of the iceberg; new studies on the domestic dog are published weekly, and new books appear regularly. There are also some very strong opinions out there on dogs, as any scan of the internet and social media will demonstrate, and some of these opinions are more fact-based than others. In recent years, through this growing body of research, our understanding of how dogs think, what dogs may or may not feel, how they communicate with us and with each other, and how best to teach them to live successfully in human households has greatly improved, in tandem with increasing understanding of dog's nutritional requirements, treating medical and behavioral problems, and how dogs age.

Figure 1.4: Our "island dog," Bea, looking concerned. (Photo: E. Grigg)

Just as my mother noted for dogs, there are also a lot of dog books in the world! Despite this apparent wealth of information, many owners of dogs, dog companions and "pet parents," and professionals who work with dogs often ask us for advice (in our roles as Certified Applied Animal Behaviorists) on working with their own dogs. When I was teaching clinical companion animal behavior to veterinary students, I had many a well-meaning (and well-educated) student come to me asking for advice on how to resolve a behavior or training question about their dog, how best to socialize their puppy, how to avoid making existing behavioral problems worse, or some similar issue. They were often overwhelmed by the conflicting information they were getting from friends, family, the media, the internet, and even other veterinarians. A very common refrain was, "I want to make this better, but I'm not sure what to do. I just want to be sure that I'm doing the right thing!"

How do we decide who to believe, when it comes to making decisions about our own dogs?

So, what is "the right thing"? As noted above, there is certainly no lack of opinions on what the "right thing" is. We believe there are two important rules of thumb for choosing an approach to our dogs. First, that whatever course of action we take is based on, and supported by, the science.

Figure 1.5: The goal of this book is to provide information helpful in ensuring that your dog lives a happy life. (Photo: ©Donna Kelliher, www.DonnaKelliher.com)

There are researchers, veterinarians, applied behaviorists, and other qualified experts[9] who have dedicated their impressive careers to studying the domestic dog, their behavior, and how they interact with their human caretakers. I want to be sure that what I am doing, and what I am recommending to others, is not just based on my opinion, but agrees with the information that these experts provide on how domestic dogs think, learn, and respond to us, and thus relies on scientifically-proven techniques for providing the best care for our canine companions. Second, and this is a more personal one, I want to be sure that my actions are always humane, ethical, and compassionate, to the best of my ability, in any given situation. Our dogs do so much for us; we owe them this, at least.

What do our dogs need to be happy? How to use this book

So, it was with those goals in mind that this book was written. We wanted to provide an easy-to-use resource for those owners who in turn want to provide the best possible lives for their dogs, have questions that need answers, and who want to be secure in the knowledge that their choices for their canine charges are based on the best available scientific information. What do our dogs really need to have a good quality of life; what do our dogs really need to be happy? And, what does "happy" mean, from a dog's perspective? There is now a significant body of research on the emotional lives of dogs: what emotions they do feel, to the best of our ability to assess this (see Box 1.2). Animals do feel emotion: the parts of the brain that produce emotion are nearly identical in humans and non-human animals; animals have evolved to possess the equipment to produce many of the same basic emotions that we experience. In addition, drugs effective in changing emotional state in humans, such as anti-anxiety drugs and anti-depressants, have very similar effects on animals, indicating that similar processes are occurring within the brains of both. But, unlike us, dogs don't have verbal language; so, what does their body language (their primary mode of communication with us and with other dogs) tell us about what emotion they are experiencing? We'll talk much more about understanding canine body language in Chapter 4. Does our dogs' quality of life affect their emotional state, and vice versa? (In a word, yes.) Dogs can feel happy, and fearful, and even angry; dogs can be optimists or pessimists; recently, researchers have begun investigating whether dogs feel grief in ways similar to the way that we experience this emotion.

Each chapter in this book will cover a different aspect of living and working with dogs, and you as the reader can decide whether to read the whole book cover-to-cover, read the chapters in order or not, or just skip to the chapters of greatest interest to you. Of course, we hope you'll read the whole thing, but you should follow the strategy that is most useful for you. Regardless of how you proceed, we do recommend that you begin with Chapter 2, as that chapter discusses the concept of quality of life and what is known about factors important to a dog's quality of life. In Chapter 2,

---

9 The question of what we (and the majority of our colleagues) consider a "qualified expert"– in addition to licensed veterinarians, Certified Applied Animal Behaviorists, and researchers in this field – will be discussed in Chapter 5 (on training), Chapter 6 (on Canine Mental Health), Chapter 9 (on choosing a veterinarian) and for examples, see the "Annotated bibliography" section at the end of this book.

> **Box 1.2. What does the science say about emotions in dogs and other non-human animals?**
>
> As of the date of this book's publication, it is generally agreed in the scientific community that many non-human species do experience the "basic" emotions (such as happiness, fear, sadness, and anger). Whether these species also experience the more "complex" emotions (such as shame, guilt, and jealousy) is still debated, although as Dr. Frans de Waal noted in 2011:
>
>> Instead of assuming that animal emotions are necessarily simple and straightforward, it is more likely that if humans and related species respond similarly under similar circumstances, the emotions behind their responses are similar, too. The latter view postulates fewer psychological changes in a relatively brief evolutionary time, hence is more parsimonious* than the assumption that unique mechanisms underpin human behavior.
>
> Dr. de Waal wrote a wonderful review of what is known about emotions in non-human animals (not just dogs) for the New York Academy of Sciences (see de Waal 2011).
>
> Emotions evolved as an important form of social communication (allowing individuals to better understand the emotions of other members of their group or species), and dogs appear to be very skilled at reading the emotions not just of other dogs, but of humans as well. Recent scientific research using functional magnetic resonance imaging (fMRI; a non-invasive tool for studying brain activity in real time); on mirror neurons (a specialized type of neuron that responds to, and reflects, actions observed in others); on empathy; and in the field of comparative neuroscience (the study of the nervous and sensory systems of animals across species) all provide support for the ability of dogs to read both canine and human emotions accurately.
>
> The comparative neuroscientist Dr. Jaak Panksepp has done groundbreaking work on emotion in non-human animals, and argues against the "relative brain size" approach to judging animals' ability to feel emotion. He notes that the "deep brain" areas such as limbic system and amygdala are shared between humans and non-humans, supporting the existence of emotions in non-human animals. Dr. Panksepp also argues eloquently against the idea that we should not study animal emotions because "we will never know what they really feel"; instead, he points out, "if emotions (aka affective states) are influencing behavioral choices, and are perhaps playing a primary role in the reward/reinforcement vs. punishment/cost decisions, studying only the observable behavior without consideration of underlying emotions is missing a 'critical scientific dimension'." For a very readable summary of Dr. Panksepp's work, see Chapter 5 in Morell 2013.
>
> *In other words, accounts for what we see in a more feasible, straightforward, relatively simple way.

we introduce the practical framework we are recommending to help you understand, assess, and improve (if necessary) the quality of life of your own dog. Each chapter will include an overview of the current state of the science in that topic area, along with the practical applications of this information (i.e., "How do I use this information in the real world, with my own dog?"). When relevant, we will discuss common misconceptions about the topic, and review what the science tells us on these specific issues. Although we do occasionally cite our own published research and experiences, our primary goal for this book is that the information and advice contained within it is supported by the latest scientific research on dogs and dog behavior. For this reason, the book is first and foremost based on a vast amount of information provided by a broad variety of pretty wonderful, brilliant, and well-respected researchers and dog experts. At the end of each chapter, therefore, we provide a list of references and recommended resources, for those of you who would like to (or perhaps I should say, have the time to) read the original research on which the chapter is based, or get more detailed information on a given topic from someone we trust to provide solid facts and excellent advice. We are learning more each day, which really does make this such a wonderful time to live and work with dogs. Additional information, useful links and content updates will also be provided as they become available.

What this book isn't

Finally, we stress that this book is not intended to be a substitute for obtaining the necessary medical care for your dog from a licensed veterinarian, or for seeking help for serious behavioral concerns from a qualified veterinary behaviorist (DACVB) or certified applied animal behaviorist (CAAB), or even for getting hands-on help when needed from an experienced, rewards-based, force-free trainer when teaching basic canine life skills. This book is also not meant to be an exhaustive, in-depth textbook covering the intricacies or controversies of the science involving domestic dogs; there are other recent books which do an excellent job of this, many of which we will list in the 'References and additional resources' section at the end of each chapter. But, it is our hope that this book will provide the information that you as fellow dog-lovers need to provide the best quality of life for your dog, from puppyhood to old age, in a user-friendly format, and as such will prove a helpful resource for dog owners everywhere. All dogs are individuals, and some have an easier time of it in life than others, but every dog deserves to be happy.

References and additional resources

American Veterinary Medical Association (AVMA). (2012) *US Pet Ownership and Demographics Sourcebook*. Available at www.AVMA.org. Accessed 9/2015.

Beck, A.M. (2014) The biology of the human-animal bond. *Animal Frontiers* 4(3): 32–36.

Beetz, A., Uvnäs-Moberg, K., Julius, H., & Kotrschal, K. (2012a). Psychosocial and psychophysiological effects of human-animal interactions: The possible role of oxytocin. *Frontiers in Psychology* 3: 234.

Beetz, A., Julius, H., Turner, D., & Kotrschal, K. (2012b). Effects of social support by a dog on stress modulation in male children with insecure attachment. *Frontiers in Psychology* 3: 352.

Berns, G. (2013) *How Dogs Love Us: A neuroscientist and his adopted dog decode the canine brain.* New York: Houghton Mifflin/Talking Dogs LLC. 248 p.

Braun, C., Stangler, T., Narveson, J., & Pettingell, S. (2009). Animal-assisted therapy as a pain relief intervention for children. *Complementary Therapies in Clinical Practice* 15: 105–109.

Cutt, J., Giles-Corti, B., Knuiman, M., & Burke, V. (2007). Dog ownership, health and physical activity: A critical review of the literature. *Health & Place* 13: 261–272.

de Waal, F. (2011). What is an animal emotion? *Annals of the New York Academy of Science* 1224: 191–206.

Endenburg, N., & Baarda, B. (1995). The role of pets in enhancing human well-being: Effects on child development. In I. Robinson (Ed.), *The Waltham Book of Human-Animal Interactions: Benefits and responsibilities.* Oxford, UK: Elsevier Science Limited. 162 p.

Friedmann, E., & Son, H. (2009). The human-companion animal bond: How humans benefit. *Veterinary Clinics of North America: Small Animal Practice* 39: 293–326.

Gácsi, M., Maros, K., Sernkvist, S., Faragó, T., & Miklósi, A. (2013). Human analogue safe haven effect of the owner: Behavioural and heart rate response to stressful social stimuli in dogs. *PLoS One* 8(3). DOI: 10.1371/journal.pone.0058475.

Grandin, T., & Johnson, C. (2009). *Animals Make Us Human: Creating the best life for animals*. New York: Houghton Mifflin Harcourt. 352 p.

Grimm, D. (2015). Dawn of the dog. *Science* 348: 274–279.

Hoffman, J. (2015). *The look of love is in the dog's eyes*. New York Times Well Pets blog, April 16, 2015. *http://well.blogs.nytimes.com/2015/04/16/the-look-of-love-is-in-the-dogs-eyes/?_r=0*. Accessed 9/2015.

Kaminski, J., & Marshall-Pescini, S. (2014). *The Social Dog: Behavior and cognition*. San Diego, CA: Academic Press. 404 p.

Kis, A., Bence, M., Lakatos, G., Pergei, E., Turcsan, B., Pluijmakers, J., ... Kubinyi, E. (2014). Oxytocin receptor gene polymorphisms are associated with human directed social behavior in dogs (*Canis familiaris*). *PLoS One* 9(1): 1–9.

Levinson, B.M., & Mallon, G.P. (1997). *Pet-oriented Child Psychotherapy* (2nd ed.). Springfield, IL: Charles C. Thomas Publishing, Ltd. 242 p.

McGreevy, P.D., & Nicholas, F.W. (1999). Some practical solutions to welfare problems in dog breeding. *Animal Welfare*. 8: 329-341.

Miklósi, A. (2015). The science of a friendship. *Scientific American* Special Issue 24(3): 15–23.

Miklósi, A. (2007). *Dog Behaviour, Evolution and Cognition*. Oxford: Oxford University Press. 276 p.

Morell, V. (2013). *Animal Wise: How we know what animals think and feel*. New York: Broadway Books. 291 p.

Nagasawa, M., Kikusui, T., Onaka, T., & Ohta, M. (2009). Dog's gaze at its owner increases owner's urinary oxytocin during social interaction. *Hormones and Behavior* 55: 434–441.

Oyama, M. A., & Serpell, J. A. (2013). General commentary: Rethinking the role of animals in human well-being. *Frontiers in Psychology* 4: 374. doi:10.3389/fpsyg.2013.00374

Payne, E., Bennett, P.C., & McGreevy, P.D. (2015). Current perspectives on attachment and bonding in the human-dog dyad. *Psychology Research and Behavior Management* 8: 71–79.

Pet Food Manufacturers' Association (2015). *Pet Population Report*. Available at www.pfma.org.uk/pet-population-2015 (accessed 6/21/2016).

Shannon, L.M., Boyko, R.H., Castelhano, M., ... Boyko, A.R. (2015). Genetic structure in village dogs reveals a Central Asian domesticated origin. *PNAS* early online edition, doi:10.1073/pnas.15162151112.

Stoeckel, L.E., Pallye, L.S., Gollub, R.L., Niemi, S.M., & Evins, A.E. (2014). Patterns of brain activation when mothers view their own child and dog: An fMRI study. *PLoS One* 9(10): e107205. October 2014.

Trut, L. (1999). Early canid domestication: The farm-fox experiment. *American Scientist* 87(2). DOI: 10.1511/1999.2.160.

Wood, L., Giles-Corti, B., & Bulsara, M. (2005). The pet connection: Pets as a conduit for social capital? *Social Science & Medicine* 61: 1159–1173.

Chapter 2

Quality of life in dogs

Photo: iStock

IN HER BOOK, *BONES WOULD RAIN FROM THE SKY*, THE AUTHOR AND DOG TRAINER Suzanne Clothier writes,

> To step into a dog's mind requires that you step into his paws and see the world through his eyes. To understand his prayers, you must look for what lights his entire being with joy, and look also for what dims that light.
>
> (Clothier 2002, p. 14)

We cannot, of course, ever really see our dog's world through his eyes; to believe we can puts us at risk of some pretty blatant anthropomorphism. Anthropomorphism is the attribution of human qualities to non-human animals. It can (and does) play a beneficial role in

Figure 2.1: Owned family dogs in St. Kitts, West Indies (photo: M. Frelinger); street dog in St. Kitts (photo: A. Loftis)

the formation and strengthening of the human–animal bond, but it can also interfere with our ability to accurately interpret our dog's behavior, particularly if we attribute human motivations to their canine behaviors (see also Bradshaw and Casey 2007). However, to me, Clothier's words can be read as a reminder that to best understand our dogs and provide them with a truly good quality of life, we need to at least attempt to see their lives from their canine point of view. We can respect our dogs by paying attention to their body language and vocalizations, and "listening" to what they are trying to communicate about the situation in which they find themselves (and more often than not, the situation into which we have placed them). If you are reading this book, then you care enough about your canine companion to want to understand them, and to give them the best quality of life possible. But, what constitutes a great quality of life, in dog terms? Dogs' lives vary so much, depending on their living situation, their physical and mental health, age, breed, and history; the beliefs and values of the human culture that surrounds them, and their role in society (companion dog, working dog, etc.). Is it possible to define a universal set of requirements for a good quality of life for domestic dogs? And, how do we know (other than in situations of obvious abuse or neglect) what an individual dog's quality of life is really like? I remember a half-serious argument I once had with a veterinarian colleague; I argued that dogs were of course better off living in a human household (vs. as street or stray dogs), with the reliable food and shelter that such a home provides. He countered by observing that many dogs in the home had little if any chance to express many natural dog behaviors, and spent much of their time alone. He suggested that stray dogs may actually have a better quality of life from the dog's perspective, with freedom to roam, mate, and sleep at will. We eventually agreed to disagree, as I still felt that the scarcity of food, disease and injury risks, and so on, of life as a stray dog outweighed any positive aspects of that life; but, his comments stayed with me, and were part of the motivation for this book. How can we give our dogs a great quality of life, from *their* perspective, as well as from ours?

How can we give our own dogs the best possible quality of life?

Given the "domestic" part of their history, of primary importance for domestic dogs may be the presence or absence of a caring human companion in their lives. Dr. Amy Marder, an expert on the behavior and welfare of dogs living in shelter situations, once said in response to an audience question about the risks of adopting a shelter dog versus acquiring a puppy from a breeder, that the only real difference between dogs in a shelter and dogs living in a home was that the dogs living in a home had a human who was committed to their care. Even for dogs living in homes, though, we all know how much and in how many ways individual human lives can vary. All these individual differences between people and households can influence a human's ability to care for a dog. So, how do we begin to talk about achieving the *best* possible quality of life (or QOL) for a domestic dog – once we figure out what this even is, from a dog's perspective?

Measuring quality of life in dogs

In the science of studying dogs, QOL in dogs is assessed in one of two ways (see also Hewson et al. 2007, for a critical review of quality of life assessments in dogs). The first is a proxy assessment of QOL, which involves asking the kinds of questions often asked in a routine veterinary visit, and tends to focus on the dog's physical well-being. Most veterinarians today would agree that the dog's mental wellness is as important as his or her physical health when evaluating QOL; it's just much more difficult to objectively measure mental wellness. Proxy information is necessary, of course, as we cannot simply ask the dog how he is doing or what she is feeling on a daily basis. Even when we can ask the subject these questions directly (such as when assessing quality of life in human subjects), we are still dealing with an abstract concept, with many possible definitions, and we are dependent on the willingness of the subject to answer questions honestly and completely. Some authors note that quality of life cannot be measured directly, even in humans. Instead, with dogs, we rely on the familiarity of the dog's owner or caretaker with the normal behavior and appearance of his or her dog, and the medical expertise and experience of the veterinarian in examining the dog and measuring certain physiological parameters. Observations of behavior or appearance that are not "normal," based on the owner's familiarity with their dog, or of exam findings that are not within the normal range of values for that measurement in healthy dogs, are generally taken as a sign that there may be a problem with the dog's physical or mental well-being. These observations can be considered fairly objective in the case of the veterinarian who has just done a physical exam on the dog, and is working with test results, diagnostic x-rays or similar. However, owner observations of their own dog can be highly subjective, and can depend on that person's background knowledge about what constitutes normal behavior in dogs, the amount of time he/she spends with the dog, even that person's personal attitudes as to what constitutes an appropriate level of care for a dog.

Another type of QOL assessment is more objective, has its origins in the science of animal welfare, and is used not just in studies of companion dogs and cats, but also of zoo animals, farm animals, and animals living in research laboratories. This type of assessment is based on quantitative measurement of observable behavior, with certain types of behavior being considered signs

> **Box 2.1. Signs of stress in dog**
>
> Dogs are individuals, and their responses to stress will vary, depending on their genetics and physiology, their past experiences, and the nature and duration of the stressor. However, common signs of chronic stress in dogs (based on studies of dogs in stressful situations such as shelters) include:
>
> Maladaptive behaviors that do not benefit, and may be harmful, to the animal, such as self-mutilation.
>
> Repetitive and stereotyped behaviors, such as spinning in circles, excessive, continuous barking or jumping.
>
> Behaviors indicating frustration (such as chewing, vocalizing) or conflicted emotions (body-shaking, paw-lifting).
>
> A lowered/fearful body posture.
>
> Activity level changes, such as a marked increase or decrease in activity; increased activity in response to stress is more commonly reported.
>
> Apparent overreaction to relatively mild stimuli (often triggering repetitive behaviors such as circling and jumping), perhaps due in part to the dog's frustration at not being able to reach and interact with the stimulus.
>
> Sources: Beerda et al. 2000; Hewson et al. 2007; Beerda et al. 1998; Hetts et al. 1992; Hubrecht et al. 1992.

of acute or chronic stress (see Box 2.1), and of physiological stress responses such as cortisol levels (a primary physiological indicator of stress in both dogs and humans) in the animal's blood, saliva, urine, or hair. These assessments also rely on the researcher's knowledge of what is "normal" in the species that they study, but can provide a more convincing, data-driven picture of how various living situations, handling, and so on, affect an animal's quality of life. This type of assessment is frequently conducted, for example, on dogs living in shelters or other types of kennel facility. Perhaps the best, most complete picture of a given animal's quality of life can be obtained by using an integrated approach, combining proxy assessment with objective measures of observable behavior and physiological stress parameters (Hewson et al. 2007).

Quality of life is a lifelong goal

In clinical veterinary practice, the question of an individual dog's QOL usually arises in the context of caring for aging or very sick dogs; caretakers want to monitor changes in the dog's QOL in order to provide the best care that they can, and to know when it may be time to let their friend go. For veterinarians, Morton (2007) outlined a detailed method for evaluating quality of life, particularly when euthanasia may be a consideration. However, QOL is a lifelong phenomenon, and should be considered throughout the life of the dog, from puppyhood through adulthood to the senior years; not just when the dog's QOL is significantly declining. But, while most of us may be able to conduct a proxy assessment of the wellness of our dog by answering questions about his or her behavior and appearance, we are unlikely to be measuring cortisol levels or videotaping our dog's behavior on a regular basis (although, I could be wrong on that last point, in the age of YouTube, Instagram and Facebook). Perhaps the most important tool in understanding what your dog enjoys or does not enjoy, what (in Clothier's words) "lights his entire being with joy" as well as "what dims that light," is the ability to accurately read his body language. Chapter 4 will cover this topic. Canine body language is communication; this is how your dog tells you how he is doing, in that moment. And fortunately, there are frameworks which provide a practical, comprehensive way for us to assess the quality of life of our dogs, goals to shoot for and areas where things could be improved, and so to work towards increasing our dogs' quality of life to the best of our ability.

Evaluating quality of life in our own dogs: the Five Freedoms

One concept that has existed in the field of farm animal welfare for some time, but has more recently been explicitly applied to companion animals,[1] is the concept of the Five Freedoms (see Box 2.2). The Five Freedoms concept was first introduced in 1965, in a UK report[2] on the welfare of livestock under intensive agricultural use; these were later formalized by the UK Farm Animal Welfare Council as a guiding principle for insuring that animals' most basic needs are met. Essentially, the Five Freedoms are meant to summarize the five most basic needs of any animal; a deficit in any one of these five areas can be a significant cause of stress, and cause a decline in well-being. Many animal welfare organizations, including dog groups, now recognize the Five Freedoms as the guiding principles for providing humane animal care. As guiding principles, the Five Freedoms are meant to serve as goals: while it may not be possible to meet and maintain all five at every possible moment, owners and caretakers should always strive to provide all five to the animals in their care, to the greatest extent possible.

The Farm Animal Welfare Council (2009) goes further to state that quality of life can be classified into three general categories as: life not worth living; life worth living; or, a good life. Providing an animal with the last of these (i.e., a good life, vs. just a "life worth living") requires "skill and conscientiousness"; in other words, knowledge of what the species requires, and a commitment to providing what is required. So, that is our goal for this book: to give you the knowledge that you, as a dedicated dog owner, need to be able to provide a good life for your dog(s), knowledge in all five of these areas – diet, housing, choosing a veterinarian, behavior, and mental well-being.

We should strive to provide all Five Freedoms to our dogs, to the greatest extent possible

The first three of the Five Freedoms are (hopefully) considered minimum requirements in this day and age when taking on the responsibility of caring for a dog (Box 2.2), although (as you'll see in the chapters on physical wellness and choosing a veterinarian) differences of opinion exist on the best ways to do this. With the last two Freedoms, however, many dog owners find themselves on much shakier ground, not through lack of willingness or affection for their dog, but rather through a lack of knowledge about how dogs think, learn, what emotions dogs experience (and how to recognize these emotions in your dog), and what they need as social animals. Often, based on the way our dog is behaving, we are aware that something is not quite right in her world (and

---

1 In 2010, the Association of Shelter Veterinarians (US) used the Five Freedoms as the basis for their impressive and comprehensive "Guidelines for Standards of Care in Animal Shelters"; see also the Grandin and Johnson 2010 reference at the end of this chapter.
2 Now known as the Bramwell Report (December 1965); available from the Farm Animal Welfare Council in the UK.

> **Box 2.2. The Five Freedoms**
>
> 1. Freedom from Hunger and Thirst (by providing ready access to fresh water and a diet to maintain full health and vigor)
>
> 2. Freedom from Discomfort (by providing an appropriate environment including shelter and a comfortable resting area)
>
> 3. Freedom from Pain, Injury, or Disease (by prevention or rapid diagnosis and treatment)
>
> 4. Freedom to Express Normal Behavior (by providing sufficient space, proper facilities, and company of the animal's own kind)
>
> 5. Freedom from Fear and Distress (by ensuring conditions and treatment which avoid mental suffering)
>
> *Source: Farm Animal Welfare Council, 2009.*

this can lead to significant stress in our human world); but, we're not sure of the best way to address or even identify the problem. In some cases, we may have been given bad advice and are now faced with the feeling that things are just getting worse, despite our well-intentioned attempts to stop an unwanted behavior, reduce our dog's fearfulness, or train better behavior. Not knowing what our dog needs can be frustrating, for both parties.

The last two Freedoms, in fact, are very important to quality of life, and the research demonstrates this again and again. As Marian Stamp Dawkins, the well-respected Oxford University professor of animal behavior, stated, "Let us not mince words: animal welfare involves the subjective feelings of animals" (Dawkins 2011). Our dogs' quality of life is influenced by how they feel, how they perceive their environment, and whether their basic emotional needs are met. Dawkins goes on to state that animals are more likely to suffer if they are not provided with resources for which they are highly motivated; so, what motivates our dogs? This is where the framework provided by the fourth and fifth Freedoms can be helpful. For example, dogs are a social species, and spending time with other dogs is important for their quality of life; this is part of the fourth Freedom, "freedom to express natural behavior". A number of studies conducted in kennel situations have learned that dogs do better, even in the stressful environment of a kennel or shelter, when housed with other dogs (rather than alone) (Hubrecht et al. 1992; Mertens and Unshelm 1996; Hetts et al.

Figure 2.2: Socializing dogs. (Photos: S. Watko, R. Hack)

1992; and Beerda et al. 1999). Group or pair housing is now recommended by numerous authors and scientists as a standard practice any time dogs are housed long-term (ASV 2010). However, our dogs are individuals, and not all dogs are happy or even comfortable interacting with a large group of dogs. We as caretakers need to watch our own dog carefully, paying attention to his or her body language, to determine whether our dog is a dog who enjoys (for example) visits to the crowded off-leash dog park, or who would rather spend time with us, or with one other dog friend. The social lives of domestic dogs also include their interactions with humans; spending time having fun with our dogs is an important part of allowing them the freedom to be dogs. We will talk a lot more about the social lives of dogs in Chapter 6, including recommendations on ensuring that your dog has the social life that he or she needs. In Chapters 6 and 8, we'll talk about giving your dog other opportunities to "just be a dog," including regular physical exercise, a stimulating and enriched environment, and perhaps involvement in any of a number of dog sports such as agility training, nose work, and more.

The fifth Freedom, freedom from fear and distress, is equally important for quality of life in dogs, and this too is supported by the science. For example, in a 2010 survey of 721 dog owners whose dogs had died, the stress of living with a fear or anxiety disorder was associated with a shorter lifespan (Dreschel 2010). There was also a tendency for dogs categorized as "well-behaved" to have longer lifespans. Sadly, given that behavior problems are a primary reason for relinquishment to a shelter and for euthanasia of otherwise healthy dogs, this is not a surprise. Fear, we now know, is at the root of many behavioral problems in domestic dogs. Temple Grandin's wonderful book, *Animals Make Us Human*, provides an in-depth discussion of what is known about the emotional needs of many species, from chickens to cheetahs to domestic dogs (with examples of problems that occur when an animal's emotional needs are not met, drawn from studies of zoo, livestock, and companion animals). Drawing from her 30 plus years of working with animals, Grandin encourages animal caretakers to focus on activating the animal's positive emotions as often as possible, and on not activating their negative emotions (of which fear is one) any more than necessary (Grandin and Johnson 2010). This advice ties in well with another aspect of considering QOL in dogs from the dog's perspective. Humans generally evaluate their own QOL by integrating how they feel about their current state with expectations of future changes, hopes and goals, and so on. Dogs' brains, on the other hand, may not be built to integrate all these abstract ideas in the same way that we do. Dogs, it is often said, live in the moment; therefore our evaluation of QOL must be an ongoing process – how happy is the dog right now? How can we maximize the happy moments in our dogs' lives, and minimize the unhappy ones?

Stress ... it's not just for humans anymore!

The mammalian stress response is a normal physiological response to a potentially threatening situation, and can be beneficial (when it occurs only briefly), in helping to get the individual safely out of the dangerous situation. However, our bodies, like those of our dogs, were not designed for living with continual or even frequently repeated stress. Chronic stress, often associated with suboptimal living conditions, has been shown to be associated with health problems in a wide range of species, including humans, ranging from reduced immune function (opening the way for disease to become established in the body), digestive problems (including vomiting and diarrhea), urinary tract issues (cats are notorious for this particular response), skin problems, and behavioral problems (including phobias and fear-based aggression). Stress has also been shown to interfere with learning in some contexts. In some situations, short-term stress can actually increase focus and learning; how many of us can still remember exactly where we were and what we were doing on that awful morning of September 11, 2001, or of July 7, 2005? Chronic stress, on the other hand, can interfere with learning new information and retrieving memories of lessons learned in the past. The relationship between stress and learning is a very complex one; for more on this and an example of research into this question, see Blackwell et al. (2010). A 2014 report by Daniel Mills, a specialist in clinical animal behavior, and his colleagues urged veterinarians to recognize the relationships between stress and animal health, to acknowledge that individual animals will have individual responses (based on the type of stress involved, and on that animal's appraisal of the situation), and to take active steps to reduce stress in order to improve the overall health of their patients. Mills et al. (2014) give a nice summary of impacts of stress on non-human animals, and recommendations for reducing stress in their lives.

An important thing to come out of studies like those described above is that dogs (like people) differ in their tolerance for, and responses to, stress; all the more reason to pay attention to what our dog's body language is telling us in any new or potentially stressful situation. This won't come as a surprise to anyone who has trained service dogs, scent detection dogs, and the like; not all dogs are suited for working in high-stress situations. Numerous training and assessment programs exist (and are continually being refined) to try and identify which dogs will do well "under fire," and which would be better suited to a peaceful life as a companion. Like humans,

Figure 2.3: Stress can affect dogs in many ways, and can often contribute to undesirable behaviors. (Photo: iStock)

dogs have different coping styles,[3] resulting in some dogs being able to handle stressful situations that other dogs cannot. I saw this regularly while working with dogs housed for long periods of time in a university teaching dog colony; some dogs quickly developed stress-related behaviors such as excessive barking, spinning, and repetitive jumping; other dogs remained calm and appeared content, seeming to be completely unfazed by life in the kennels.

Mills et al. (2014) recommend that when necessary, veterinarians conduct a "stress audit" (see Box 2.3), in which they systematically evaluate various aspects of the dog's home life, taking into account the individual dog's apparent responses to these various aspects, and looking in particular for known "risk factors" – situations known to be stressful for many dogs. Stress, after all, is cumulative, and the more risk factors present, the more likely we are to see negative impacts on the dog. Figure 2.4 illustrates some of the most common ways that well-meaning owners inadvertently put their dogs in stressful situations.

Practical applications: the "Five Freedoms test"

The Five Freedoms give people who live with dogs a starting point from which to think about their own dog's quality of life, and a framework for considering where to make improvements, if necessary. If we consider each and all of these five areas, and whether our dog truly has each of these Freedoms, we can begin to feel confident in the quality of the care we are providing (or, identify areas where changes may need to be made). Dogs whose basic needs are met are more likely to live longer lives, with fewer behavioral problems along the way. When it is simply not possible to allow a dog to express normal dog behavior (such as roaming the neighborhood off leash and alone, an activity that is unsafe, and illegal in many areas), what can we do to provide an alternate activity that meets this emotional need? Can we think of ways to make small changes in our household in order to give our dogs choices, providing them with some measure of control over their environment? When it comes to recommendations for action, each chapter of this book will deal with a specific area and provide the most up-to-date recommendations, supported by the science (such as the most effective, most humane training methods to use, in Chapter 5; how to enrich your dog's mental and social worlds, in Chapters 6 and 7; what to feed your dog, in Chapter 8; and so on).

So, thinking back to my old argument with my colleague about quality of life for companion versus street dogs – what can the Five Freedoms tell us about the relative quality of life of these two groups of dogs? I lived and worked for four years on the Caribbean island of St. Kitts, where there was a substantial population of street dogs, and I worked with many dogs adopted as companions from this population. Using those "island dogs" as an example, as well as (given my experience with many companion dogs in the US) what I think a typical (i.e., neither the best nor worst possible) living situation is for a companion dog, I've conducted a quick "Five Freedoms Test." In this test, I've detailed how I feel that the Five Freedoms are met or not met

---

[3] Professor J.M. Koolhaas and colleagues have done a lot of interesting work on coping styles in animals, and the relationship between the two primary coping styles, proactive versus reactive, and vulnerability to stress; see Koolhaas et al. 1999, referenced at the end of this chapter, for an example.

Some dogs really do seem to love children, others tolerate them; but for many dogs (especially those not raised around children), children (with their erratic movements, loud voices, etc.) can be very scary. This is also a fairly high-risk situation, for both the child and (should anything bad happen) the dog and his owners. For this reason, most dog experts recommend never leaving a dog alone with a child, and always being watchful for signs of discomfort or stress in the dog's body language and behavior. If the dog appears stressed, immediately remove him (or the child) to a safe distance.

Some dogs are true social butterflies, and thrive in the high energy, often chaotic environment of the dog park; for these dogs, dog parks are a great way to meet their social needs and provide them with physical exercise. Other dogs, however, find this situation stressful, either due to a dislike of large groups of dogs in general, or due to the actions of one or more specific dogs present on a given day. We owe it to our dogs to pay attention to them while at the dog park (just as we would be attentive to a child at a crowded playground), and intercede when necessary to ensure a peaceful, enjoyable experience for everyone.

A fenced yard is a wonderful thing to have when living with a dog. However, if the dog is left unattended for hours on end in a fenced yard, and able to hear and/or see countless passersby (canine or otherwise) throughout the day, significant frustration can develop. This can develop into what is known as "barrier aggression," familiar to anyone who has seen a dog rush a fence as another dog passes by, barking as though the world is about to end. Allowing this behavior to persist is not kind to the dog, nor to the passersby. In situations where more than one dog is enclosed in the yard, this intense frustration can sometimes result in displaced aggression, where one dog (unable to reach the dog or person on the other side of the fence) will turn on his companion.

Just as some dogs are very friendly towards children, many dogs will happily and eagerly greet strangers on the street or visiting their home. However, for many dogs, the approach of an unfamiliar human or dog is a scary, potentially threatening situation. These dogs usually tell us how they feel through their body language, by cringing, backing away, or growling. If we ignore these early signals of discomfort, the reactions to strangers can progress to barking, or sometimes even lunging, snapping, or biting. Forcing a fearful dog to remain in place while a stranger approaches is not kind to the dog, and will rarely help the dog become comfortable in this situation. In many cases, it will make the behavior worse (and may put the approaching stranger at risk of defensive aggression). We will talk about alternate ways to reduce fear of strangers in Chapter 6, on Canine Mental Health.

Figure 2.4: Common stressful situations for companion dogs.

Box 2.3. The stress audit for dogs

The stress audit looks at aspects of the dog's daily routine, such as:

How much time does the dog spend home alone (and, what is there for him to do while he is home alone)? A sterile, unenriched environment has been associated with stress in many animals (imagine being in solitary confinement for days and weeks on end, with nothing to do and nothing to read). For more on the importance of environmental enrichment, as well as tips for how to provide it for your own dogs, see Chapter 6.

How predictable is her daily routine? Many dogs do not do well with unpredictable schedules, and can become stressed.

Does the dog have any way to control his own environment, particularly in the presence of a stressful person or situation? For example, does the dog have a "safe place" (such as a crate, a dog door to the yard, or a bed in a quiet room) that he can retreat to when he is uncomfortable or wishes to escape the situation for a while? In my own experience, this can be particularly important for dogs not comfortable with strangers, when unfamiliar guests visit. The ability to control, at least to some degree, our environment is an important aspect of mental health; in humans, feelings of being unable to control our lives are commonly associated with stress, and conditions like clinical depression. In dogs, the inability to exercise control over their environment is believed to contribute to common canine behavior issues like barrier frustration and leash reactivity (in that case, the inability to reach the person or dog on the other side of the fence, or beyond the reach of the leash).

If expectations are placed on the dog's behavior, is support provided to help her meet these expectations, for example through rewards-based training of acceptable alternate behaviors? This one reminds me of my own dogs growing up; my parents had a strict rule of "no dogs on the furniture," but my parents also taught night classes on many weekday evenings. As soon as they left in the evening to teach, my siblings and I would not only allow the dogs on the furniture, but would happily encourage them to climb up with us; we loved having them share the couch with us while we read or watched television. My parents were increasingly frustrated with the dogs' seeming inability to learn and obey the rule. In truth, the inconsistent message the dogs were getting would be confusing (and, I realize now to my chagrin, stressful) for anyone. Fortunately, my parents are the forgiving sort. A consistent message is important for teaching and learning; if we give our dogs contradictory or confusing information, we are much more likely to have problems getting them to behave as we would like.

Adapted from Mills et al. 2014.

for these two groups of dogs. I've scored the level of each Freedom on a scale of 1 (never or rarely met) to 5 (always or almost always met).[4] My co-author, based on her many years of working with companion and shelter dogs, completed the same exercise; the scores shown in Table 2.1 are the averages of our two scores for each category. Because we are both "data nerds," I've also included the average score for all Five Freedoms, for each group of dogs.

Of course, unlike most of the information that this book presents, scoring on this exercise is

---

4 Note that many users of the Five Freedoms use a binary approach, i.e., either the Freedom *is* met, or *is not* met, for that animal in that situation (vs. the gradient from 1 to 5 that we used here).

Table 2.1 The "Five Freedoms Test" as a way to assess quality of life in dogs. Rating scale for each Freedom is from 1 (never or rarely met) to 5 (always or almost always met).

Freedom:	Street dogs	Companion dogs
from hunger and thirst	2	5
from discomfort	3	5
from pain, injury, and disease	1.5	4
to express normal behavior	5	3
from fear and distress	2	3.5
Average score:	2.7	3.9

based on our "expert opinion," versus any systematic scientific investigation. However, in almost all categories, our scores were exactly the same, even though we completed the exercise without knowing how the other had scored each category. When we did vary, it was only by one point (as may be obvious by the 1.5 and 3.5 ratings in the figure). And looking at the average score, I still believe that I was right in that argument years ago – companion dogs do have a better life, overall, than street or stray dogs. However, the average score was not as different as the average dog owner might think it would be. If our goal is to get scores as close to 5 as we can, then there are areas where we can do better. The last two Freedoms (four and five) are where this is most evident. I do believe that there are some dog owners out there whose scores on meeting these last two Freedoms would be much higher, perhaps even 5s. But, if you perform this same exercise as objectively as possible for your own dog (considering your dog's point of view as of primary importance in how you score each category) and you don't get 5s across the board, we hope that the chapters that follow will help you raise your scores.

Next steps: moving beyond the Five Freedoms

The Five Freedoms concept has been, and continues to be, a strong force for good in the animal welfare arena. However, a number of experts in animal welfare are now encouraging people who live and work with animals to think beyond this initial framework. The Five Freedoms, after all, primarily focus on minimizing negative influences on welfare (for instance, animals should be "free from hunger and thirst"), with only one ("freedom to express natural behaviors") focusing on increasing positive influences on welfare. For example, Professor David Mellor of Massey University notes that some negative experiences can never realistically be completely eliminated, merely temporarily neutralized (hunger and thirst, for example). In fact, temporarily experiencing these factors is essential for eliciting behaviors on which the survival of the animal depends (Mellor 2016). If you have ever returned hungry, thirsty, and tired after a long (but rewarding) hike, bike ride, workout or the like, you know how satisfying it is to eat, drink, and rest after such physical exertion. The negative experiences provide motivation for, and pleasant anticipation of, the activities which function to reduce these feelings. As we suggested

Figure 2.5: Sometimes, a little bit of stress is unavoidable – such as at bath time! (Photo: S. Watko)

earlier, and as Dr. Temple Grandin noted in her work (Grandin and Johnson 2010), we should strive to minimize negative influences on our dog's welfare, while enhancing the positive influences, to the greatest extent possible, each and every day. In the wording of the Five Freedoms, Dr. Mellor suggests, "Freedom from" should perhaps be replaced by "As free as possible from" (Mellor 2016). A balance is needed, in our dogs' lives, as in our own, between the unavoidable (but hopefully, temporary) negative feelings and experiences, and the more positive, enduring ones. As Dr. Hannah Buchanan-Smith and her colleagues at the University of Stirling note, the ability to exercise some control over negative welfare states is also important to quality of life. Imagine how you'd feel at the end of that long hike if you discovered that there was no food or water available at that time, or for the foreseeable future! In some cases, even the *perception* of having control over their environment can be beneficial for captive and companion animals (as, realistically, we cannot always grant them real, unrestricted control over their environment; Herrelko et al. 2015). Dr. Buchanan-Smith[5] also recommends that we think about these issues from a practical standpoint: if questioned, what *evidence* could we provide to demonstrate that we are meeting these goals, whether the Five Freedoms or beyond, for our canine companions?

---

5 Dr. Buchanan-Smith and her colleague Sabrina Brando have put together a program advocating for the addition of choice, complexity, and control to animals' lives to improve welfare; more on their work can be found at their project website, *http://www.247animalwelfare.eu/*.

Dogs in our lives and homes: the rules are changing

Our evolving relationship with our canine companions, and the accompanying responsibilities and expectations for care of these companions, are becoming evident in the legal sphere as well. If we look back over the history of how dogs (and cats) were perceived in human society, things have changed drastically over the years. In his very readable book, *Citizen Canine*, scientist and writer David Grimm documents these changes, and notes the progression in the US from dogs and cats being considered less than property (with no penalties for their theft or killing), to a form of property (with owners free to do with them as they liked), to the current state of affairs, in which states enact strict anti-cruelty laws and courts grant damages for mental suffering to plaintiffs whose pet has been killed. The increasing body of evidence linking cruelty and violence towards animals with cruelty and violence towards people (e.g., Degenhardt 2005, Randour 2004) adds weight and urgency to the passing of these anti-cruelty laws. After the devastation of Hurricane Katrina in the southeastern US, and all the attention focused on the plight of individuals and families trying to save their pets, even refusing to evacuate to safety without them, the US federal government enacted the PETS (Pets Evacuation and Transportation Standards) Act. The PETS Act of 2006 requires that state and local emergency preparedness operational plans "address the needs of individuals with household pets and service animals following a major disaster or emergency," including providing rescue, shelter, and care for companion animals.[6] More and more, our companion animals are being recognized as part of our family.

Setting realistic expectations is important for our canine companions

Given their status as part of the family, with better health care and nutrition than perhaps ever before, in some respects QOL for many dogs is at an all-time high. However, this view of dogs as part of the family has a downside for some dogs. It is a wonderful thing that so many of us consider our companion animals as part of our family, but if we extend this to the point where we expect them to behave like *human* members of our family, *without teaching them how we wish them to behave*, then we do them a disservice. This can extend to assigning motives to their behavior that may not exist, such as guilt, attempts to dominate human household members, and similar; we'll discuss these issues more in the chapters on canine cognition, and humane training. The way in which we treat our dogs is influenced by our beliefs of how our dogs perceive their world; if we over or underestimate their sensory or cognitive abilities, or their ability to feel certain emotions, we may inadvertently be putting them in situations detrimental to their welfare (Bradshaw and Casey 2007). Similarly, just because we enjoy a given activity, does not mean that our dogs will ever share this enjoyment; we have a tendency to project our own emotions onto our dogs. Our dogs are individuals, with their own likes and dislikes (and, above all, they are dogs, not humans. Repeatedly and unnecessarily exposing them to situations that

---

6 the AVMA has a helpful FAQs web page on the PETS Act, at https://www.avma.org/KB/Resources/Reference/disaster/Pages/PETS-Act-FAQ.aspx.

they find stressful is not a particularly nice way to treat a member of the family. As an example of this, when I lived and worked on an island I spent much of what free time I had at the beach, and I cannot count the number of times I watched otherwise well-meaning students dragging their dogs (often young puppies) into the water with them, cheerfully proclaiming that they were teaching their dogs not to be afraid of the ocean. The dogs were not physically harmed in any of the incidents that I saw, but it was clear from their body language that the vast majority were highly fearful, struggling to return to shore, and almost certainly were not learning to love the ocean through this exercise. Other students were savvier, teaching their dogs how fun the ocean could be by combining the water with a game of fetch, throwing the ball into the shallows for the dog to retrieve, and gradually throwing the ball further out in the water. When we adopted our own island puppy, she was fearful of many things (including the ocean). We never forced her to "confront her fears"; rather, she learned through daily beach walks (her favorite activity) and at her own pace to approach and explore the edge of the sea, eventually following our older dog into the shallows. By the time we left the island she would happily race along splashing through the shallow waves (but, she was never comfortable going out of her depth … and that's OK).

Quite simply, we may be asking more of them than they, as dogs, are able to provide. When I was growing up, we were taught to leave the family dogs alone while they were eating or sleeping, because they are dogs, and sometimes dogs bite if upset or caught unawares.[7] Nowadays, "resource (or food) guarding" is considered by many owners as an abnormal behavior in their dog (it is not abnormal behavior for a dog, although it may be unacceptable and unsafe in a family dog). What is normal for a dog has not changed much in the recent past; however, what we consider "acceptable" in our dogs has changed quite a bit, and will likely continue to change. And, as the trainer Jean Donaldson notes in her book, *The Culture Clash*, when the dog's way of life conflicts with human rules and regulations, this often has serious, even fatal consequences for the dog. This concept goes back to the earlier observation that providing an animal with best quality of life requires knowledge of that species – what is normal behavior for them, and what are their basic needs. John Bradshaw puts it bluntly in his wonderful book, *Dog Sense*:

> The new, unrealistic standards to which many humans hold their dogs have arisen from one of several fundamental misconceptions about what dogs are and what they have been designed to do. We must come to better understand their needs and their nature if their niche in human society is not to diminish.
>
> (Bradshaw 2011, p. xviii)

An important question here, as noted back in the stress audit (Box 2.3), is this: if expectations are placed on the dog's behavior (and sometimes, these expectations are necessary if the dog is to live safely and comfortably within a human household), is support provided to help the dog meet these expectations? Providing support, in most cases, means providing humane training of acceptable, alternate behaviors. We'll talk about how to do this, and about effective and humane training methods for your dog, in Chapter 5.

---

7 Our family dogs never did bite, as it happens, but perhaps because we rarely put them in situations where they might have been motivated to bite.

This relationship between a dog and his human companions, this human–animal bond, is vitally important to his quality of life; the existence and strength of this bond has been found to play a role in everything from success in training, to successful adoption of shelter dogs into homes, to successful resolution of behavioral problems. Just the fact that you are reading this book, and perhaps other books like this, is a good sign for your commitment to providing your canine companion with the best life possible, and to sharing your home with a happy dog.

Take-home messages

1. Currently, quality of life in dogs is assessed in a number of ways, ranging from owner surveys to behavioral analyses to measurement of physiological parameters. Assessing our own dog's quality of life requires an accurate understanding of, and attention to, their body language, which can help us to see any situation from the dog's perspective.
2. The Five Freedoms can provide owners and caretakers with a framework for assessing their own dog's quality of life, as well as goals for optimal care.
3. Providing the best quality of life for our dogs requires knowledge of their needs, their normal behavior, and how they perceive their world. What do our dogs like to do, and how can we find ways to enable them to do these enjoyable activities as often as possible, given our other commitments and responsibilities?
4. Expectations placed on a dog's behavior should be realistic, given their nature as dogs; if we wish to modify their essential canine tendencies, any expectations of alternate behaviors should be supported with humane training.
5. Each chapter that follows will discuss an aspect of dog's lives, to provide the knowledge needed to help meet the Five Freedoms for your dog.

References and additional resources

Association of Shelter Veterinarians (ASV) (2010). *The Association of Shelter Veterinarians' Guidelines for Standards of Care in Animal Shelters.* Newbery, S. et al., editors. 67 p.

Beerda, B., Schilder, M.B.H., Bernadina, W., Van Hoof, J.A.R.A.M., De Vries, H.W., & Mol, J.A. (1999). Chronic stress in dogs subjected to social and spatial restriction. II. Hormonal and immunological responses. *Physiology & Behavior* 66: 243–254.

Beerda, B., Schilder, M.B.H., Van Hoof, J.A.R.A.M., De Vries, H.W., & Mol, J.A. (1998). Behavioural, saliva cortisol and heart rate responses to different types of stimuli in dogs. *Applied Animal Behaviour Science* 58: 365–381.

Beerda, B., Schilder, M.B.H., Van Hoof, J.A.R.A.M., De Vries, H.W., & Mol, J.A. (2000). Behavioural and hormonal indicators of enduring environmental stress in dogs. *Animal Welfare* 9: 49–62.

Blackwell, E.-J., Bodnariu, A., Tyson, J., Bradshaw, J., & Casey, R. (2010). Rapid shaping of behaviour associated with high urinary cortisol in domestic dogs. *Applied Animal Behaviour Science* 124: 113–120.

Bradshaw, J.W.S. (2011). *Dog Sense: How the new science of dog behavior can make you a better friend to your pet.* New York: Basic Books. 324 p.

Bradshaw, J.W.S., & Casey, R.A. (2007). Anthropomorphism and anthropocentrism as influences in the quality of life of companion animals. *Animal Welfare* 16(S): 149–154.

CAAB Chats, with Patricia Hetts and Dan Estep (2015). *Pampered Pets or Stressed Out, Overburdened Companions,* with guests Amy Marder, Patricia McConnell, and Steve Zawistowski. October 2015; available online at: http://caabchats.com/pampered-pets/ (accessed November 2015).

Clothier, S. (2002). *Bones Would Rain from the Sky: Deepening our relationships with dogs.* New York: Grand Central Publishing/Hachette Book Group. 305 p.

Dawkins, M. S. (2011). From an animal's point of view: Motivation, fitness, and animal welfare. *Behavioral and Brain Sciences* 13: 1.

Degenhardt, B. (2005). *Statistical Summary of Offenders Charged with Crimes against Companion Animals July 2001–July 2005.* Report from the Chicago Police Department.

Donaldson, J. (2005). *The Culture Clash* (2nd ed.). Oakland, CA: James & Kenneth Publishers. 203 p.

Dreschel, N. (2010). The effects of fear and anxiety on health and lifespan in pet dogs. *Applied Animal Behaviour Science* 125: 157–162.

Farm Animal Welfare Council. (2009). *Five Freedoms.* Available at: http://www.fawc.org.uk/freedoms.htm.

Grandin, T., & Johnson, C. (2010). *Animals Make Us Human: Creating the best life for animals.* New York: Mariner Books. 352 p.

Grimm, D. (2014). *Citizen Canine: Our evolving relationship with cats and dogs.* New York: Public Affairs. 336 p.

Herrelko, E.S, Buchanan-Smith, H., & Vick, S-J. (2015). Perception of available space during chimpanzee introductions: Number of accessible areas is more important than enclosure size. *Zoo Biology* 34(5): 397405.

Hetts, S., Clark, J., Calpin, J., Arnold, C., & Mateo, J. (1992). Influence of housing conditions on beagle behaviour. *Applied Animal Behaviour Science* 34: 137–155.

Hewson, C., Hiby, E.F., & Bradshaw, J.W.S. (2007). Assessing quality of life in companion and kennelled dogs: A critical review. *Animal Welfare* 16(5): 89–95.

Hubrecht, R., Serpell, J., & Poole, T. (1992). Correlates of pen size and housing conditions on the behaviour of kennelled dogs. *Applied Animal Behaviour Science* 34: 365–383.

Koolhaas, J., Korte, S., De Boer, S., Van der Vegt, B., Van Reenen, C., Hopster, H., ... Blokhuis, H. (1999). Coping styles in animals: Current status in behavior and stress-physiology. *Neuroscience & Biobehavioral Reviews* 23: 925–935.

Mellor, D.J. (2016) Updating Animal Welfare Thinking: Moving beyond the "Five Freedoms" towards "A Life Worth Living". *Animals* 6: 1–20, doi:10.3390/ani6030021.

Mertens, P., & Unshelm, J. (1996). Effects of group and individual housing on the behavior of kennelled dogs in animal shelters. *Anthrozoos* 9(1): 40–50.

Mills, D., Karagiannis, C., & Zulch. (2014). Stress – its effects on health and behavior: A guide for practitioners. *Veterinary Clinics of North America: Small Animal Practice* 44: 525–541.

Morton, D.B. (2007) A hypothetical strategy for the objective evaluation of animal well-being and quality of life using a dog model. *Animal Welfare* 16(S): 75–81.

Randour, M. L. (2004). *Including animal cruelty as a factor in assessing risk and designing interventions.* Conference Proceedings, Persistently Safe Schools, The National Conference of the Hamilton Fish Institute on School and Community Violence, Washington, DC.

Chapter 3

Canine cognition: how your dog sees the world

Photo: iStock

I've seen a look in dogs' eyes, a quickly vanishing look of amazed contempt, and I am convinced that basically dogs think humans are nuts.
John Steinbeck (1962) Travels with Charley; in search of America

IT MAY BE HARD TO TELL IF DOGS THINK WE ARE CRAZY OR NOT, BUT WE DO KNOW THAT dogs can deduce a lot from us, their human companions. Most of us that have spent any time around dogs have witnessed the half-cocked head and quizzical expressions that many dogs possess; but what underlies that expression? Can they really tell what we are thinking, and question our intentions, as they seem to do with that look? Dogs can read humans amazingly well, even in comparison to our closest relative, the chimpanzee. Over time, dogs have evolved specialized cognitive and social skills through their close association with us that allow for social interactions between two very different species (Miklósi 2009). These interactions can even be viewed as an analog of human behavior (Topál et al. 2009). This is a rather unique relationship between two species that live and work in close association and which influence each other. By exploring this relationship we can determine just what dogs have learned from living with humans and just what capacities they have for complex communication, cognition, and emotion.

Studies show that we are important to dogs

Humans are as important to the lives of dogs as they are to us. There is evidence to suggest that domestic dogs have attachments to humans and display behaviors similar to those seen in human infants (Tomasello and Kaminski 2009). These attachments underlie the specialized communicative abilities present in the modern day dog that allow for not only information exchange between two species, but for the creation of an enduring social bond. If you think that your dog can tell how you are feeling, you are right! Studies show that dogs have the ability to recognize human expressions and decipher emotional content in both facial expressions and vocalizations (Albuquerque et al. 2016). This finding is the first reported evidence that dogs attend to the emotional responses of a species other than their own. In addition to reading our emotions, dogs use information from us in other ways. Dogs follow human gestures to find hidden food better than any other mammalian species, and they even use eye contact to determine if the information is intended for them (Kaminski et al. 2012). Dogs can tell when we are paying attention to them and when we are not. Dogs also use communication from us to gain information about unfamiliar situations (Merola et al. 2012), a phenomenon called social referencing. They show differences in how they respond to a new situation based on our reaction to the situation, and on the level of familiarity they have with us (Merola et al. 2012). Dogs really are man's (and woman's) best friend, and in this chapter we will explore the unique skills of social cognition that dogs use to better relate to us. We will investigate further what dogs understand about humans and what this means for our relationship with them. We will look at the concept of guilt in dogs and explore it as an example of how our perception of canine cognition and behavior influences our interactions with dogs. Lastly, we will discuss dog cognition and how it impacts dogs' quality of life; knowing what we now know about canine cognition, how do we contribute to the life of a happy dog?

Why do dogs communicate with us? The evolution of social competence in dogs

Dog caregivers know that dogs have special skills for communicating with us. As we will detail throughout this chapter, dogs excel in their ability to share information across species. Why is it that dogs in particular are so sophisticated in this area? In many tests of social cognitive ability, canines routinely outperform most primates. How did dogs become our closest confidant, more so than our closest relatives, the great apes? Domestic dogs are unusual in that they typically live in multispecies groups and it is thus advantageous for them to acquire certain social skills to improve their interactions with humans. The term "social competence" has been applied to dogs' interactions with humans. Social competence is the ability to get along with people; know what is expected for social interactions, such as making eye contact; "reading" people's facial expressions and gestures; recognizing emotions in others; and being able to communicate effectively with people. Social competence, in this case, is the idea that dogs have the skills necessary for successful social interactions with humans. Specifically, this means that dogs may have the ability to produce appropriate behavior in social interactions with people and thereby meet the expectations of human caregivers. Given that humans are biased towards animals with an enhanced interest in interacting with us, it is likely that this prosocial behavior has been selected for in domestic dogs. This was a gradual process over the course of canine evolution, in which the wolf progenitor species with more human-like prosocial behavior and compatible skills (i.e., that were better able to tolerate and interact with humans) prospered. Humans recognized these prosocial behaviors and the advantages of having a canine social companion. Humans favored these individuals by provisioning and protecting individuals with these characteristics, which in turn increased the selection pressure on dog social behaviors.

Dogs have proven themselves capable of learning many extraordinary tasks to assist humans in our work, in keeping us safe, and even in our daily lives. They are a highly intelligent social species, but researchers are currently trying to explore further the extent of social competence in dogs. How have dogs evolved to get along so well with people?

There are a few theories on how the dog evolved to have such profound social communicative abilities with humans. One theory, the *by-product* hypothesis, is that the process of domestication favored less fearful animals that tolerated the presence of humans, and thus fostered the integration of dogs into human groups and the development of social cognition (Hare and Tomasello 2005). Alternatively, the *adaption* hypothesis contends that these changes were not a result of selection for one specific trait, like a lack of fearfulness, but that canine social skills were honed by small genetic changes in various aspects of social behavior, such as attachment and cooperation (Topál et al. 2009). Lastly, others argue that this ability comes from learning (about how to best survive and thrive among humans) and development of skills that occurs during the dog's lifetime, rather than evolutionary adaptations (Udell et al. 2008). Whatever the cause, the result of this process was an evolutionarily novel interspecific social competence in dogs, a remarkable phenomenon that allowed for a wide range of relationships with humans (Miklósi and Topál 2013). But, just how much do dogs really understand about human social interactions? The next sections of this chapter will summarize some of the ways in which researchers have attempted to answer this question.

The studies that began the explosion of interest in canine cognition

The information on dogs' social cognitive abilities that ignited the escalation in dog cognition research across the globe was the finding that dogs respond to human ostensive cues (pointing, head nod, eye gaze). This was the first evidence that dogs could recognize human communication as informative. Dogs have been shown to use human communicative gestures more successfully than any other non-human animal (Kaminski and Nitzschner 2013). To investigate this, many studies have used what is called an object-choice paradigm. This is a test in which the human experimenter hides food under one of multiple containers outside of the dog's view (or smell). The human demonstrator then gives the dog a social cue (a pointing gesture, head nod, or eye gaze) in the direction of the hidden food. Dogs are very good at reading these cues, especially the pointing cue, and there is some evidence that dogs can even distinguish when the cue is intentional. Dogs monitor us and use our cues, but it is more difficult to determine if they understand our motivations, versus simply following our cues. To determine what dogs know about us, we need to understand how they pay attention to us and just what they can decode about our mental states.

Attention and perspective

Dogs know when we are paying attention to them and may understand our perspective. If you think that your dog will not capitalize on the delicious food sitting on the counter when you are distracted by your email or text messages, think again! Dogs are able to determine when you are attending to them and when you are not, and they often modify their behavior depending on your attentional state. In one revealing study, dogs were told not to take a piece of food, after which the experimenter either kept their eyes open and directed on the food, closed their eyes, feigned distraction by a computer game, or turned her back on the food. The study dogs ate less food when the human was paying attention to them, compared to all the other situations (Call et al. 2003). Dogs will also obey commands faster when the person giving the command is facing them and in sight (Gácsi et al. 2004). These findings have been replicated with numerous other studies and they all provide evidence that dogs know when you are paying attention to them, and that they can capitalize on that information.

Dogs may be more limited in understanding the human perspective, although there is some evidence that dogs will take advantage of a situation in which humans cannot see what is going on. When dogs are challenged with forbidden food, in which they can see the food but a large barrier obstructs the human's view, dogs will take the food more often than when the human can clearly see the forbidden food (Bräuer 2014). In another study using barriers to test dogs' visual perspective taking, experimenters placed a toy either within the person's visual field or not. The dog was then asked to bring the toy. Dogs consistently brought the toy that was within the visual field of the person, rather than one that could not be seen by the person (i.e., when the toy was behind an opaque barrier or behind the person's back; Kaminski et al. 2009). Both of these studies provide evidence that dogs are able to take some information about the human visual perspective and use it. However, it is uncertain how much information dogs really understand about human perspective, that is, about just how much we know in any given situation. In other words, dogs may be able to tell what we can *see* at any given point in time, but not what we *know* (based on our previous experience, for example) about the situation. One study found that dogs can differentiate between information from an informed person (i.e., a person who has seen where the food is hidden, in the presence of the dog) versus an uninformed person about the location of hidden food, but evidence for this ability is weak and other studies have not been able to replicate these findings (Cooper et al. 2003). For instance, in the second component of the study with toys and barriers mentioned previously (Kaminski et al. 2009), dogs were present when experimenters viewed toys being hidden behind opaque barriers. However, the dogs then did not retrieve those toys that the experimenters *had* seen placed any more often than they retrieved the toys experimenters *had not* seen placed (suggesting that the dogs were not taking into account what the human had or had not seen in the immediate past).

To further assess the question if dogs can determine what we know from past experiences, researchers used an apparatus with a person at one end, and a window at the other end where dogs were placed at the start of the trail (from this initial vantage point, the dogs could clearly see the person in the apparatus; Bräuer et al. 2013). The dogs were then shown a forbidden food being placed in a tunnel that ran horizontally through the middle of the apparatus (between the dog's initial viewing window and the human observer, and in view of both parties), and which the dog could potentially retrieve from the tunnel using his paw. Half of the tunnel was transparent (meaning that the person could see the dog's paw in action if he chose to retrieve the food using the transparent side) and the other half of the tunnel was opaque (meaning the human could not see the dog's paw if he chose to retrieve the food using the opaque side). Importantly, once the dog moved to either side (opaque or transparent) of the food tunnel to retrieve the food, he could no longer see the human. Researchers found that dogs that could see the person watching from their initial vantage point, but then moved to a location outside of the human's view to take the hidden food from the tunnel, did not hide their approach to the tunnel. In addition, they showed no preference for the opaque side of the tunnel (even though the person could potentially see their paw from the transparent side of the tunnel). This leads to the conclusion that once the person was out of sight of the dog, the dog did not take their perspective into account (as the researchers put it, "dogs seemed to conclude, 'if I do not see her, then she does not see me'"; Bräuer 2014, p. 300).

In the same study, Bräuer and colleagues (2013) tested to see if dogs were sensitive to what humans could hear. They found that dogs prefer to approach forbidden food silently if a person was present, even if they could not see the person while they took the forbidden food from the tunnel, and they did not show the same preference for silence when a person was not present. This indicates that dogs may not be able to take into account what humans have seen in the past, and what information humans retain from those experiences (Bräuer et al. 2103; Bräuer 2014). However, the same conclusion cannot be made in knowing what people can hear. As Bräuer (2014) points out, this may be due to the different ways that visual and auditory stimuli are propagated; just because I can't see the dog behind the barrier, doesn't mean I can't hear him when he makes noise, given that I am in the same room with him. As Bräuer (2014) notes, dogs may conclude, "When I hear the noise, then the other hears it too" (Bräuer 2014, p. 302). This study suggests that dogs are relying primarily on what they themselves can see and hear at any given time, but further work will be needed to parse out exactly why discrepancies were found between the two sensory modalities.

In some challenges dogs seem capable of understanding what we know and in others they do not. At the very least, we know that dogs do pay attention to us and can tell what we can see, and maybe they even know what we know. Future studies will help to determine if dogs are simply exceptional readers of human social communication signals in the moment, or if they can in fact use information from us in more complex ways (and perhaps, ways in which individual dogs differ in these skills, and the reasons behind these differences).

Attachment

> **Box 3.1. How your dog views you (or the human–dog relationship models)**
>
> 1. *Dominance*–submission or lupomorph: Relationship is based on human as unquestioned leaders based on wolf society. This was the predominant model for dog training but has been found to be an inappropriate model given recent findings about wolf and intraspecific dog behavior. Free-ranging dogs are semi-solitary and do not generally conform to strict dominance hierarchies.
>
> 2. *Parent–infant*: Dogs form infant-like bonds with humans, and humans in turn provide caregiving similar to that given to human infants. Humans use "baby talk" and maintain attention similar to that used with an infant. In studies of attachment using the strange situation test, dogs respond to caregivers similarly to ways in which human infants respond to their parents.
>
> 3. *Friendship*: Described as a close affiliative bond or as a relationship where affiliate behaviors outweigh aggressive ones. However, the dog–human relationship is not symmetrical or egalitarian and is primarily shaped by human needs and desires.
>
> Sources: Prato-Previde et al. 2003; Goodall 1986.
>
> For more detailed information see:
> The Social Dog: Behaviour and cognition, edited by Kaminski and Marshall-Pescini.

If you are interested enough in dogs that you purchased a book on the concept of happiness in dogs, you no doubt have experienced feelings of attachment to a dog at least at one point in your life. There is evidence to support the idea that the relationship that develops between the dog and caregiver meets the criteria for attachment as defined in human–infant studies. In studies using the strange situation test, originally developed to test strength of attachment in human children, dogs were challenged by being placed in an unfamiliar environment with their human caretaker. An unfamiliar person then entered the room. The dog was then separated from their human caregiver, and later reunited (Prato-Previde and Valsecchi 2014). Dogs showed attachment behaviors such as approaching the caregiver when stressed, using the caregiver as a safe/secure base for exploration, and exuberantly reuniting with the caregiver after the separation. Overall, results from these tests support the idea that dogs form real attachment bonds to their human caregivers (Ainsworth 1969).

There is consensus that dogs do in fact form such attachments, but there is less agreement on what form the attachment takes. There are a few models of the human–dog relationship, including the dominance–submission model, the parent–infant model and the friendship model (Box 3.1). No one model seems to perfectly define the relationship dogs have with their human caregivers, but that does not mean that these attachments are not impactful on the life of a happy dog. Clearly dogs can become stressed when the caregiver is absent, and show reliance on the caregiver when faced with new challenges.

Dogs trust your opinion: social referencing

Behave how you wish your dog to behave. That doesn't mean that your dog should act like a human, but it does mean that your dog will tend to accept the things that you accept. Dogs use human reactions to determine how they should react to a situation. This is called social referencing, defined as using another's emotional reaction to an unfamiliar object or event to determine what one's own emotional or behavioral response should be. Imagine holding your friend's hand in the haunted house at the funfair: if she isn't afraid, you aren't, but the moment she jumps, you are afraid. Dogs pick up cues from people the same way. In one study of dogs' reactivity to a novel object, dogs behaved differentially based on the response from a human (Merola et al. 2012). When a familiar person (caregiver) gave positive feedback to the dog (smiling and using a happy voice) while looking alternatively at a novel object (a fan) and then to their dog, the dogs explored more and approached the fan in less time than dogs that received a negative message from their caregivers. Familiarity mattered, too; when a familiar person gave the positive feedback, the majority of dogs (70 percent of dogs tested) made more attempts to look to their caregivers for information about the object. Only 60 percent of dogs tested with an unfamiliar person looked to the stranger for information about the fan. These results are similar to those seen in human infants.

Familiarity matters!

In the study described above, researchers found that the effectiveness of the message was decreased when a stranger delivered the message. Although dogs receiving the information from a stranger (positive or negative messages) looked to the stranger for information, they did not do so as often as they looked to the (familiar) caregiver when the caregiver delivered the message. They also did not use the information received from strangers in the same way they used information gathered from the caregivers. Unlike dogs receiving information from their caregivers, dogs receiving information from strangers showed no difference in the amount of time it took them to approach and interact with the novel object. Interestingly, dogs receiving a negative message from the familiar caregiver spent more time in the area close to the door (i.e., close to the seated caregiver and away from the fan), exhibiting more immobile behavior and looking more often to the seated caregiver. This behavior is also similar to that seen in human infants when faced with similar stressors. Taken together, these results suggest dogs were most likely sensitive to the emotional expression of the stranger, but the way that they *changed their behavior* was more dependent on

Figure 3.1: Dogs use eye contact in communicating with us. (Photo: K.D. Setiabudi)

their relationship with the informant. When caregivers were the informants and gave positive information, the dogs spent more time with the unfamiliar object, but they did not do so when given positive information from a stranger. Dogs in the study demonstrated that they do use social referencing but the information that they use is also dependent on who is delivering the information. Dogs look to their familiar human companions for guidance and direction, in many situations. In short, our dogs trust us to steer them in the right direction.

We also know that dogs can recognize familiar faces. In a discrimination test between the heads of images of the heads of caregivers or other familiar people, dogs were able to choose the face of the caregiver. They were less able to perform this task when parts of the visage were shielded and only the central part of the face (eyes, nose, mouth) was exposed (Huber et al. 2013). Functional magnetic resonance imaging (fMRI) studies in dogs have given insight into dogs' abilities for reading our faces (as well as human emotions). Through these studies we have discovered that dogs show brain activation when presented with human faces without training (Cuaya et al. 2016). Dogs also have brain areas that show activation in response to human voices similar to the regions human use for processing communication.

How dogs communicate with us

Dogs are superb companions, and one reason for this is their demonstrated ability to communicate with us. We've talked so far about what information dogs gather *from* us, but part of social competence in dogs is demonstrated by their ability to communicate *to* us. Dogs can convey information to people as well as read our signals to them. Dogs can signal the location of a hidden object by using "showing behaviors," like alternating their gaze from the object to their human companion repeatedly until the target is acquired. An example of this is when a ball-crazed dog looks to you and then to the ball that has rolled outside of the dog's reach as if to say, "See that ball? Get it please." Interestingly, there are limits on what dogs will convey to us. Dogs display these types of signals only when the target item is interesting to the dog (Kaminski et al. 2011). They do not show these behaviors for targets that are only interesting to the human. So, either they lack the motivation to retrieve an object solely for the human's benefit, or (unless specifically trained to retrieve) they do not have the capability to understand that the human is interested in the item (only whether or not they themselves are interested in it).

Do dogs understand our intentions?

There is conflicting evidence about what dogs are able to understand about our intentions. Dogs are able to distinguish between intentional pointing to a source of food versus unintentional pointing, for instance as when checking the time on a wristwatch. In a study using the object–choice paradigm, humans used either purposeful pointing to the location of hidden food (intentional pointing) or used a movement that mimicked pointing where the demonstrator extended the arm and pointed a finger while checking her watch (unintentional pointing). Dogs followed the intentional movement more than the unintentional movement (Kaminski et al. 2012). Nonetheless, researchers are not entirely certain if dogs are actually identifying and understanding our intentions, or if they are simply responding to other subtle physical cues. Our dogs appear to be paying attention to our eye contact and body language, but does this demonstrate their understanding of the deeper meaning of our ostensive cues? Dogs can follow the pointing gesture when the human's eyes are closed or when the back is turned but not as reliably as when the informant makes eye contact, suggesting that eye contact is an important element in dog–human communication. In the study of the intentional finger point versus unintentional watch-checking motions (Kaminski et al. 2012), dogs may not have understood the nature of the point and only used eye contact to find the hidden food. When the experimenter delivered the intentional finger point, she also looked alternatively to the dog and then to the hidden food. Conversely, when the experimenter gave an unintentional point while checking her watch, she looked alternatively to a clock on the wall and then the hidden food. This may indicate that the dog is relying on the eye contact to know that the person is meaning to share information (Kaminski et al. 2012).

There are limits in dogs' understanding of human communication. Dogs cannot follow the information if it is delivered to a third party (another person), rather than directly. This may also give credence to the idea that eye contact and dogs' extraordinary ability to read human body language are the key factors in determining how dogs respond to our cues. Think about how this might apply to your dog's success rate during training, or when responding to your requests or cues out in the real world! This may be why many successful trainers train their dogs to focus on them, before attempting to teach any other, more complex cues.

As we previously stated, dogs can tell when we are paying attention to them, and may often modify their behavior according to whether or not anyone is watching (something humans do, too). But, are dogs capable of complex mental abilities similar to those of the human mind? Whether or not dogs possess "theory of mind" is a current subject of interest in canine cognitive research. Theory of mind is defined as the ability to acknowledge and attribute mental states (beliefs, intentions, desires, knowledge, etc.) to oneself and others, and to understand that others have beliefs, desires, intentions, and perspectives that are different from one's own (Premack and Woodruff 1978). As noted above, dogs have shown that they are aware of human attention (or lack thereof), as well as the attention of other dogs. Horowitz (2008) looked at attention between dogs during play and concluded that, "these dogs showed attention to, and acted to manipulate, a feature of other dogs that mediates their ability to respond: which feature in human interaction is called 'attention'." Attention is considered a precursor to theory of mind, and lends support to the argument for dogs possessing this ability. Other researchers such as Brian Hare and Marc Bekoff have investigated aspects of canine cognition that support this argument, as well. If we accept that dogs experience "secondary emotions" such as jealousy, in addition to the "basic emotions" that they do possess,[1] this would also strongly support theory of mind in dogs. There is still some debate in the scientific community about which species (dogs included) possess theory of mind, and to what degree (dogs "pass" some tests, but seem to fail others; see Udell et al. 2011), but

---

1 For more on what is known about canine emotions, see Chapter 5, on training a happy dog, and Chapter 6, on mental wellness in dogs.

the bulk of the research on dogs suggests that they do indeed possess this skill, at the very least in some rudimentary canine form, and possibly at a much higher level.

We have evidence that dogs have superb abilities to use information from us. Studies like these give weight to the adaption hypothesis of canine domestication, described earlier in this chapter and stating that dogs have a specialized skill set associated with human-given directives. In other words, dogs have developed a special skill set, such as responding to eye contact for cues, for activities that humans have used them for (rather than the byproduct hypothesis in which they would have developed a broader range of skills, such as understanding our intentions, associated with domestication).

Reading our emotions: does my dog really know what I'm feeling?

Knowing that dogs are attentive to people, let us explore just what information about us dogs can decode. In humans, the face is the primary conveyor of emotion; can dogs decipher human facial expressions? One of the most exciting recent findings is that dogs can do this, and do respond to emotions of humans. In a study to determine how dogs perceive human emotions, dogs were presented with images of three emotional facial expressions (threatening, pleasant, and neutral) of other dogs and of humans, while researchers recorded the patterns of their eye gaze (Somppi et al. 2016). They found that dogs use multiple facial features of humans such as the eyes, midface and mouth to determine emotional meanings, rather than just one feature

Figure 3.2: Dogs use our facial features to respond to our emotional states. (Photo: Adobe Stock)

such as the eyes. In addition, and most interestingly, dogs responded rapidly and with avoidance behavior to the human threatening facial expression, whereas they responded to the threatening expression of dogs with increased attention only. Dogs appeared to deem angry human faces as threatening or aversive, and they used avoidance to dispel the perceived threat. Also, as in human fMRI studies, the regions dogs use for processing human voices respond more strongly to positive vocalizations; no brain areas responded more for negative vocalizations (Andics et al. 2014). This suggests that dogs and humans process these types of cues similarly. This is evidence that dogs can extract emotion information from human voices, as do we. Future studies such as these may change our perceptions of just what is possible in our canine companions; perhaps they might just have rich emotional lives similar to our own.

So the next time you come home tired and frustrated from a long day at work only to find your beloved pup has shredded your favorite shoe, think of how he may perceive your body language, tone of voice, and facial expressions. With this in mind, decide, does he really know that what he did was wrong? Or does he simply know that you are angry, and is trying to avoid a confrontation?

But, do you really mean it?

Dogs can also determine the *validity* of a human expression. In a study of dogs presented with images of human faces displaying emotions and a recording of an emotional vocalization, dogs paid attention longer to the image if the emotion expressed by the face in the image matched the emotion suggested by the type of vocalization. For instance, when a happy expression was paired with a playful voice, dogs attended to the image for a longer period compared to an image in which the facial expression did not match the vocalization (such as an angry expression with a playful voice). To avoid the possibility that the dogs being tested had simply learned which human words signaled trouble, vocalizations were given in a language unfamiliar to the dogs (Portuguese, in this case; Albuquerque et al. 2016). This means that dogs had to extract the information from one modality, such as vision, and understand the emotional content to determine if it matched the emotional content of the other modality, sound. This finding suggests that dogs have cognitive capacities not previously found outside of primates. Furthermore, extracting and integrating information from unfamiliar words demonstrates abilities not previously found outside of humans. This makes dogs truly exceptional at communicating with their people.

Synchronization between dogs and humans

We have discussed how dogs are sensitive to us and seem to change their behavior in response to our cues, and even respond to situations based on our responses. These observations have led a number of researchers to investigate whether dogs will synchronize their behavior with that of a familiar person. Synchrony is defined in the behavioral sciences as a phenomenon in which one individual matches the behavior of another (one simple example is when someone touches their own face and you touch yours almost automatically). This matching of behavior can form the basis for affiliation and emotional bonding. Synchrony has not previously been found to exist

Figure 3.3: Dogs are very responsive to cues, but do dogs match their behavior to ours? (Photo: K.D. Setiabudi)

between two different species of animals, but dogs may serve as a great model for studying its existence. We know that dogs use information gained from watching how people solve a task to then solve that task themselves. In a test in which dogs had to navigate a V-shaped fence (called a detour test) in order to receive a toy or food reward, dogs that were previously unsuccessful at the task were able to gain success after watching a human participant navigate the detour to obtain the reward (observational learning; Pongrácz et al. 2001). We also know that dogs are capable of social referencing (Merola et al. 2012). Taken as a group, these findings support the concept that dogs are responding to visual cues from humans and may be matching their behavior. For true synchrony to be possible, dogs would need to be capable of more than just social referencing or observational learning; they would need a broader understanding of our emotions and intentions. More work is required to conclusively demonstrate whether dogs are capable of understanding our emotions and matching our behavior, but given what we know thus far about dogs, they are a great model for studying whether synchronization between species is possible.

The green-eyed monster: do dogs experience jealousy?

Now that we have reviewed the evidence for the rudimentary elements involved in emotion (the ability to distinguish positive from negative emotions, for example), let us look at a more complex emotion: jealousy. Is your dog jealous (in the way that we feel jealousy), or is he guarding his most valuable resource: you? Jealousy was once believed to be a uniquely human trait because it requires complex emotions and cognitive abilities. Jealousy was thought to only be possible with beings that had a sense of self and the ability to reflect on one's self and others (i.e., that possessed theory of mind). Jealousy normally attributed to humans involves not only changing one's behavior after being treated unfairly, but also changing one's behavior when witnessing others receiving differential treatment (for instance, when another dog is offered a greater reward). In one study in which two dogs (the subject and a control dog) were asked for the behavior "sit," the subject dogs received a consistent amount of reward and the amount given to the control dog varied (either the same amount as the subject dog, from the "fair trainer," or over-rewarding or under-rewarding the control dog, from the "unfair trainer"; Horowitz 2012). Following work with both trainers, the dogs were allowed to choose which trainer to approach on their own: the fair or unfair one. When control dogs were under-rewarded, subject dogs chose both the unfair trainer and the fair trainer an equal number of times. In the over-rewarding trial, however, the dogs chose the unfair trainer. This result suggests that dogs were less concerned with fairness or jealousy and more driven by which trainer was (at least potentially) providing more food. As Horowitz (2012) titled her article summarizing that research, "Fair is Fine, but More is Better …."

However, there is another possible, primordial form of jealousy-like behavior that that has been noted in human infants (Harris and Prouvost 2014). This form of jealousy is one in which a dependent offspring fights for parental resources, and can be triggered without complex reflections on the self or others. Instead, this form is based purely on competition for parental attention. Using a model designed to test for jealousy in human infants, researchers found that dogs exhibited behaviors consistent with jealousy, such as snapping at the object of the jealousy, getting closer to the caretaker, pushing the object of jealousy away, or touching the owner in response to the caretaker paying attention to the object (stuffed) dog. Interestingly, dogs did not display these behaviors in response to the caregivers showing attention to other objects that did not resemble another dog (Harris and Prouvost 2014). Taken together, dogs may not be capable of the complex form of jealousy we experience as humans but they may possess this emotion in a simpler form. This behavior in dogs may represent evolution of a complex social skill once thought to only exist in humans, but more work is needed to delineate this emotional ability in dogs.

Word learning

The great divide once thought to separate human cognitive ability from that of other animals is language. Dog owners often boast about their dog's amazing ability to communicate with them. We have already discussed how dogs read us and use our actions and body language to communicate, but can they also learn our spoken language? Studies of a few amazing dogs may provide evidence that dogs have evolved to understand our language, at least to some degree. Rico is a Border collie who has a vocabulary of more than 200 words (Kaminski et al. 2004). Rico has been taught words associated with certain objects like toys, and can discriminate which specific item is being requested. In addition, Rico can learn new words in a single trial. For instance, when researchers tested Rico, they presented a novel object and provided the name of the object. They

Figure 3.4: A few Border collies (like the one pictured) have shown extraordinary abilities in word learning. (Photo: Adobe Stock)

then placed the object along with seven other familiar items behind a barrier and gave Rico the command to retrieve the new item. Rico was able to reliably retrieve this object. Even a month after the first trial, Rico retained the name he had learned. His abilities are similar to that of a human child tested with the same criteria. While his abilities are undeniably impressive, it is not clear if Rico really understands the referential information (that the word refers to a category and that object is within that category) or if it is just a more rudimentary form of association learning (pairing the object with the label). For example, when someone says, "fetch the sock, Rico," does Rico know that a sock is an item of clothing that resembles the shape of a human foot? Or has he just learned that the word "sock" is associated with the (otherwise meaningless) item presented to him? The researchers in this study propose that the mechanism for this type of language learning, is "fast mapping" (rapidly learning a new word by contrasting it with a previously learned word). Fast mapping may be occurring but is perhaps a function of simpler mechanisms than previously thought. Fast mapping was generally thought to require the ability to understand that an object belonged to a category of objects and so it was easily defined and assimilated into language. Dogs can appreciate that items have labels, they can match labels to objects, and they can store this knowledge in memory; but, they may not understand that words refer to categories of objects and use that information to learn words for new objects. Evidence from dog studies such as these may demonstrate that fast mapping is occurring and that it is a more straightforward process than once believed.

In a second case of word learning, Chaser, a Border collie, has demonstrated the ability to use a vocabulary of over 1,000 words. Chaser learned the names of 1,022 objects and was able to reliably retrieve these objects, as well as newly named, novel objects. This ability suggests that Chaser and Rico were aware that words refer to objects out of view, showing that they possess a referential understanding of the names, an ability normally attributed only to human children (Pilley and Reid 2011). Although both dogs appear to use referential information, Tempelmann and colleagues (2014) developed a study to further address the question of how Rico, Chaser, or any other dog, actually understands the word–object relation. They tested to see if what these dogs are displaying is in fact language acquisition through referential behavior by a human speaker (as these dogs' behavior would suggest), or if there is a simpler mechanism such as association based on the simultaneous presentation of the object with the label. They wanted to ascertain if dogs could pair labels to objects when the objects/labels were not presented together in the same space and time. To test this, they placed dogs in a situation in which caregivers gave referential cues to teach the names of objects (the items were hidden in a bucket but the human tester pointed and gazed toward the item and spoke the name of the item). Out of four dogs tested, only one dog was able to create word associations when the object was not visible during the learning phase. One dog was able to show some ability to pair the word with the hidden object using cues from his owner, but as the authors point out, they could not rule out a simpler form of learning called local enhancement (an individual is drawn to a location because they observed another individual at or near the location). Rico and Chaser (Kaminski et al. 2004; Pilley and Reid 2011) were seemingly capable of this referential word learning but this ability may push the bounds of dog cognitive ability. Given that there are only three dogs with the documented capacity for this type of language learning, and that human infants far surpass this ability in the first few years of life, dogs are limited by comparison, but they are unique among non-human animals in this ability to fast map (at least as far as we know). Future studies will help better illuminate just how dogs use our language, and whether their ability to understand words is a matter of simple associative learning or true use of language as the performance of these dogs suggest.

Guilty or not guilty; that is the question

Despite dogs' remarkable abilities to read and understand human communication, they may not always be quite as cognitively competent as their owners assume. Most people that I know that live with dogs are adamant that they have seen their dogs looking guilty. Caregivers often claim that their dogs know when they have performed a misdeed such as stealing forbidden food or breaking the house-training rules. This "guilty look" is one in which the dog will be still, pull back his ears, lower himself close to the ground, wag his tail rapidly, and avert his gaze, appearing sheepish. The dog seems to "know" he has misbehaved and cannot hide this knowledge. This assumption of guilt often causes caregivers to punish their pet by scolding, banishing, or worse (and, to feel justified in doing so, as the dog clearly "knew that what he did was wrong"). This can even lead to an apparent loss of trust, a feeling of personal attack on the part of the caregiver, or a general deterioration of the relationship between the caregiver and the dog.

But, despite how much we know dogs do know, are the dogs in such situations really showing us *guilt*? Recently, researchers have looked into this phenomenon to determine if dogs do in fact display guilt. In 2009, Alexandra Horowitz tested this phenomenon in dogs, with the help of their humans. Dogs were tested in caregivers' homes, using food treats. The treat was placed on the floor in front of the dog, dogs were given a cue by the caregiver to not eat the treat, and then caregivers were asked to leave the room. After the caregivers left the room, the dogs were either allowed to eat the treat or the experimenter removed the treat (thus making it impossible for the dogs to eat the treat). Experimenters replaced the food, or did not, and had the caregivers re-enter the room. Caregivers were instructed to either reprimand or greet the dog upon return regardless of the remaining food. Horowitz found that dogs that were reprimanded responded with guilty behaviors, *whether or not they had eaten the forbidden food*. Dogs that were greeted (even when they had in fact eaten the forbidden food) displayed no guilty behaviors. In fact, dogs that did not eat the food but were reprimanded showed the highest number of guilty type behaviors. This is evidence that what the dogs were displaying was not the complex emotion of guilt, but rather that they were responding to the cues given by their caregiver, regardless of their own behavior (Horowitz 2009).

Furthermore, later studies showed that even when dogs had eaten a forbidden treat, but caregivers had no evidence of the misdeed, caregivers were unable to determine if the dog had eaten the treat or not based on the dog's greeting behavior (Ostojić et al. 2015). In the absence of scolding, the "guilty look" was not seen, suggesting that the "guilty look" behaviors depend primarily on the caregiver's concurrent behavior (such as scolding or other negative reactions). In some cases, caregivers claim that they see this guilty behavior in dogs *prior to* their discovery of the misdeed. Given the evidence we have about guilt, it is more likely that the dog has built an association with items and the environment and the human's emotional reaction (such as learning that when there is a shredded shoe on the floor, the caregiver becomes really threatening). Dogs may react to the current association (without understanding that something they did at an earlier point in time resulted in the angry response from their human) by showing fear and appeasement gestures, which we then interpret as guilt. As noted above, studies have shown that people are not very accurate at determining whether the dog had really done the misdeed or not. Dogs are keen readers of our body language, and sometimes this is misinterpreted as a deeper level of knowledge or understanding. This should not diminish our respect for the incredible social intelligence of our canine companions. However, the next time you think your dog is manipulating you, think again; there is likely another, simpler (and more inherently canine) explanation for your dog's behavior.

I have heard numerous reports from friends and clients about how their dogs really dislike them or another person because the dog steals their socks or defecates on their possessions. It seems to be a strangely human trait that we default to thinking our dog is waiting to betray us or is plotting against us, rather than thinking that they are simpler beings and that there is a more guileless explanation for their behavior. Dogs know what is familiar and what brings comfort. Scent is a very important sense for dogs, and dogs choose items with familiar scent (Box 3.2) to comfort them; this is most likely why they collect your socks. They may eliminate somewhere simply because of the familiar and comforting scent of their caregiver (or in some cases, because they can smell evidence of a previous housetraining 'mistake' in this location), which our human noses cannot detect. In any case, it is often a failure on our part when dogs do such misdeeds. It is our responsibility as caregivers to provide the appropriate training and management of the environment to prevent these misdeeds. Before punishing your dog because "he knows better," think how you could have prevented the inappropriate behavior. Have you done everything possible to properly housetrain your pup (bearing in mind that every dog is different, and speed of housetraining varies between dogs)?[2] If he is eliminating in the house, most likely it is a failure of, or lapse in, training, or even a medical issue. If he is stealing items, perhaps he is not comfortable being left alone with the run of the house. If he does not like his crate, perhaps you need to revisit crate training. We'll talk more about issues like these in the chapters on training and behavioral problems. Do not take these things as a rejection by your dog, but do look to see where your training might have failed or look to a professional (whether it be a veterinarian, who

---

2 For more tips on housetraining, see Chapter 5 on training a happy dog.

Box 3.2. What the nose knows: dogs and scent

1. *What can dogs detect*: Dogs sense odors and use odors that we cannot detect. Minute amounts of an odor are detected and recognized due to the extraordinary sensitivity of the dog's nose. Dogs have up to 300 million olfactory cells compared to 5 million in humans. The brain area for olfaction in dogs is 40 times greater than in humans. Their sense of smell is 10,000 to 100,000 times more acute than our own and they can discriminate odors in parts per trillion.

2. *How dogs process scent*: Dogs have muzzles lined with specialized tissue which contains cells for scent detection as well as an organ called the vomeronasal organ for detection of pheromones. This allows dogs to detect and discriminate things that evade our detection such as changes in our cells and hormones. Dogs can recognize human scent from cells left behind on shoes, clothes, bedding, and other items of ours.

3. *How dogs' scent ability benefits us*: Tracking: dogs help to detect from as little as five steps the direction a person has travelled.

Drug and bomb detection: dogs are used to discriminate smells of bombs and drugs by our police and security teams to help prevent threats of violence.

Disease detection: dogs have been trained to identify the smells of certain types of cancers and have been used to accurately detect new cancer growths in people.

Seizure alert: dogs can detect the behavioral changes and possibly pheromone changes that signal an oncoming seizure and can be trained to alert.

For more detailed information see:

Hepper and Wells 2005; Domestic Dog Behavior and Cognition, Horowitz 2014.

---

can rule out medical issues, or a certified trainer or behaviorist to help you with training). Dogs have evolved to be with us and are meant to be our companions, depending on us for food and support; they haven't evolved to betray or attempt to dominate us. Love them and do what you can to create a happy life for your dog.

Practical applications: applying what we know about dog cognition to make a happy life for our dogs

Our actions greatly impact the quality of life of our dogs. How can we use the current information available on dog social cognition to help our dogs live happier lives?

Based on the scientific research now available, we know that our emotions can impact those of our dogs; we can cause our dogs to look guilty, to cower in fear from us based on our perceptions of their transgressions, even when they have committed no wrongdoing. We also know that dogs form strong attachments to us, and their welfare is dependent on us. Dogs may be capable of more complex abilities and emotions such as jealousy, and may even be able to synchronize their emotional behavior to ours. In fact, our dogs are dependent on us for much more than their physical well-being; we are largely responsible for their mental and emotional well-being as well. Think of the Five Freedoms (discussed in Chapter 2). In order to have a happy dog, we should strive for a life where all Five Freedoms are met for our dogs, whenever possible. This includes providing them with the freedom to express normal, natural canine behaviors (the fourth Freedom), whenever it is appropriate and safe to do so, and with a life free of fear and distress (the fifth Freedom), to the best of our abilities. To accomplish this, we need to be aware of how our actions reflect our emotions when we are interacting with our beloved companions.

Think of what your dog *knows* and *does not know*, based on the science, and then give him the benefit of the doubt. Show him the right way of doing things, go back to your training basics and make sure that you are not assuming knowledge and skills that he just has not yet mastered. Consider how your own behavior may be influencing that of your dog: are you acting fearful when another dog approaches, and thus teaching your dog to be afraid, or worse, to be defensively aggressive? We know that our dogs look to us for guidance and direction; be sure you are providing the information and training that your dog needs to succeed in this world.

We are still learning about just how much our dogs understand us, but we can take what we do know into account during our daily interactions with our dogs. We can respect the abilities that they do have, try not to expect more than they are able to provide, and help them to live the happiest lives possible. As your dog's loving caregiver, you are in the best possible position to achieve this goal for your dog.

Take-home messages

1. Dogs have social cognitive abilities that surpass those of many other animal species, and some abilities that surpass even those of our closest living relatives in the animal kingdom: the non-human primates.
2. Dogs form genuine attachments to humans, and these attachments are the basis for their unique ability to understand our language: both our body language and to some degree, our verbal language.
3. Dogs can read referential signals in humans, and gather and use information from humans.
4. Dogs may be capable of some human-like qualities such as learning words, and may even exhibit synchrony and jealousy, but they do have their limits, such as is the case with guilt. Do not assume that dogs are acting in response to emotions such as guilt and punish your dog because of this attribution; studies do not support that dogs feel guilt; at least not as we currently define it.
5. On the other hand, consider that dogs are certainly capable of more, mentally, than we had previously thought. They can recognize our faces and read our emotions; they look to us for information to help them navigate their worlds, and for support when they feel threatened. Research into the cognitive abilities of dogs continues, and will no doubt answer many of the remaining questions in the near future.

References and additional resources

Ainsworth, M.D.S. (1969). Object relations, emotional dependency and attachment; A theoretical view of a mother-infant relationship. *Child Development* 40: 969–1025.

Albuquerque, N., Guo, K., Wilkinson, A., et al. (2016). Dogs recognize dog and human emotions. *Biology Letters* 12.

Andics, A., Gácsi, M., Farago ,T., Kis, A., & Miklósi, A. (2014). Voice-sensitive regions in the dog and human brain are revealed by comparative fMRI. *Current Biology* 24: 574–578.

Berns, G.S., Brooks, A. W., & Spivak, M. (2015). Scent of the familiar: An fMRI study of canine brain responses to familiar and unfamiliar human and dog odors. *Behavioural Processes* 110: 37–46.

Bräuer, J. (2014). What dogs understand about humans. In J. Kaminski and S. Marshall-Pescini (Eds.), *The Social Dog: Behavior and cognition*. San Diego, CA: Academic Press.

Bräuer, J., Keckeisen, M., Pitsch, A., Kaminski, J., Call, J., & Tomasello, M. (2013). Domestic dogs conceal auditory but not visual information from others. *Animal Cognition* 16: 351–359.

Call, J., Bräuer, K., Kaminski, J., & Tomasello, M. (2003). Domestic dogs (*Canis familiaris*) are sensitive to the attentional state of humans. *Journal of Comparative Psychology* 120: 257–263.

Cooper, J., Ashton, C., Bishop, S., et al. (2003). Clever hounds: Social cognition in the domestic dog (*Canis familiaris*). *Applied Animal Behaviour Science* 81: 229–244.

Cuaya, L.V., Hernández-Pérez, R., & Concha, L. (2016). Our faces in the dog's brain: Functional imaging reveals temporal cortex activation during perception of human faces. *PLoS ONE* 11: 1–13.

Gácsi, M., Miklósi, Á., Varga, O., et al. (2004). Are readers of our face readers of our minds? Dogs (*canis familiaris*) show situation dependent recognition of human's attention. *Animal Cognition* 13: 311–323.

Goodall, J. (1986). *The Chimpanzees of Gombe*. Cambridge, MA: Belknap Press. 673 p.

Grassmann, S., Kaminski, J., & Tomasello, M. (2012). How two word-trained dogs integrate pointing and naming. *Animal Cognition* 15: 657–665.

Hare, B., & Hare, V. (2013). *The Genius of Dogs: How dogs are smarter than you think*. New York, NY: Plume Books. 367 p.

Hare, B., & Tomasello, M. (2005). Human-like social skills in dogs? *Trends in Cognitive Sciences* 9: 439–444.

Harris, C.R., & Prouvost, C. (2014). Jealousy in dogs. *PLoS ONE* 9: e94597.

Hepper, P.G., & Wells, D.L. (2005). How many footsteps do dogs need to determine the direction of an odour trail? *Chemical Senses* 30: 291–298.

Horowitz, A. (2008). Attention to attention in domestic dog (*Canis familiaris*) dyadic play. *Animal Cognition* 12: 107–118.

Horowitz, A. (2009). *Inside of a Dog: What dogs see, smell, and know*. New York, NY: Scribner.

Horowitz, A. (2012). Fair is fine, but more is better: Limits to inequity aversion in the domestic dog. *Social Justice Research* 25: 195–212.

Horowitz, A. (Ed.) (2014). Domestic Dog Cognition and Behavior: The scientific study of *Canis familiaris*. New York, NY: Springer.

Horowitz, A., & Hetch, J. (2014). Looking at dogs: Moving from anthropocentrism. In A. Horowitz (Ed.), *Domestic Dog Cognition and Behavior*. New York, NY: Springer. pp. 201–220.

Huber, L., Anaïs., R., Scaf., B., & Virányi, Z. (2013). Discrimination of human faces in dogs. *Leaning and Motivation* 44: 258–269.

Kaminski, J., & Marshall-Pescini, S. (2014). *The Social Dog: Behaviour and cognition*. San Diego, CA: Academic Press.

Kaminski, J., & Nitzschner, M. (2013). Do dogs get the point? A review of dog-human communication ability. *Learning and Motivation* 44: 294–302.

Kaminski, J., Call, J., & Fischer, J. (2004). Word learning in a domestic dog: Evidence for "fast mapping". *Science* 304: 1682.

Kaminski, J., Bräuer, J., Call, J., & Tomasello, M. (2009). Domestic dogs are sensitive to a human's perspective. *Behaviour* 146: 979–988.

Kaminski, J., Neumann, M., & Bräuer, J., et al. (2011). Dogs *canis familiaris*, communicate with humans to request but not to inform. *Animal Behaviour* 82: 651–658.

Kaminski, J., Schultz, L., & Tomasello, M. (2012). How dogs know when communication is intended for them. *Developmental Science* 15: 222–232.

Merola, I., Prato-Previde, E., & Marshall-Pescini, S. (2012). Dogs social referencing towards owners and strangers. *PLos ONE* 7: e47653.

Miklósi, A. (2009). Evolutionary approach to communication between humans and dogs. *Veterinary Research Communications* 33: 53–59.

Miklósi, Á, & Topál, J. (2013). What does it take to become 'best friends'? Evolutionary changes in canine social competence. *Trends in Cognitive Sciences* 17: 287–294.

Ostojić, L., Tkalčić, M., & Clayton, N.S. (2015). Are owners' reports of their dogs' 'guilty look' influenced by the dogs' action and evidence of the misdeed? *Behavioural Processes* 111: 97–100.

Pilley, J.W., & Reid, A.K. (2011). Border collie comprehends object names as verbal referents. *Behavioural Processes* 86: 184–195.

Pongrácz, P., Miklósi, Á., Kubinyi, E., et al. (2001). Social learning in dogs. I. The effect of a human demonstrator on the performance of dogs (*Canis familiaris*) in a detour task. *Animal Behaviour* 62: 1109–1117.

Prato-Previde, E., Valsecchi, P. (2014). The immaterial cord: The dog-human attachment bond. In J. Kaminski, J., and S. Marshall-Pescini (Eds.), *The Social Dog: Behavior and cognition*. San Diego, CA: Academic Press.

Prato-Previde, E., Custance, D.M., Speizio, C., & Sabatini, F. (2003). Is the dog-human relationship an attachment bond? (An observational study using Ainsworth's strange situation). *Behaviour* 40: 225–254.

Premack, D. G., & Woodruff, G. (1978). Does the chimpanzee have a theory of mind? *Behavioral and Brain Sciences* 1: 515–526.

Somppi, S., Törnqvist, H., Kujala, M.V., et al. (2016). Dogs evaluate threatening facial expressions by their biological validity – Evidence from gazing patterns. *PLoS ONE* 11: e0143047.

Soproni, K., Miklósi, Á., Topál, J., & Csányi, V. (2001). Comprehension of human communicative signs in pet dogs. *Journal of Comparative Psychology* 115: 122–126.

Soproni, K., Miklósi, Á., Topál, J., Csányi, V. (2002). Dogs' responsiveness to human pointing gestures. *Journal of Comparative Psychology* 116: 27–34.

Tempelmann, S., Kaminski, J., & Tomasello, M. (2014). Do domestic dogs learn words based on humans' referential behaviour? *PLoS ONE* 9: e91014.

Tomasello, T., & Kaminski, J. (2009). Like infant, like dog. *Science* 325: 1213–1214.

Topál, J., Miklósi, Á., Csányi, V., & Dóka, A. (1998). Attachment behaviour in dogs (Canis familiaris): A new application of Ainsworth's (1969) strange situation test. *Journal of Comparative Psychology* 112: 219–229.

Topál, J., Miklósi, A., Gácsi, M. et al. (2009). The dog as a model for understanding human social behavior. *Advances in the Study of Animal Behavior* 39: 71–116.

Udell, M.A.R., Dorey, N., & Wynne, C.D.L. (2008). Wolves outperform dogs in following human social cues. *Animal Behaviour* 76: 1767–1773.

Udell, M. A., Dorey, N.R., & Wynne, C.D. (2011). Can your dog read your mind?: Understanding the causes of canine perspective taking. *Learning & Behavior* 39: 289–302.

Vilá, C., Savolainen, P., Maldonado, J.E. et al. (1997). Multiple and ancient origins of the domestic dog. *Science* 76: 1687–1689.

Chapter 4

Reading your dog

Photo: iStock

CHARLES DARWIN WROTE IN HIS GROUNDBREAKING WORK, *THE EXPRESSION OF THE Emotions in Man and Animals*, that, "the behavioral expression of emotion in animals is not only elusive, but is complicated by our own human reaction to its display" (Darwin 1872). Dogs communicate differently than humans, and our dogs' emotions may be somewhat elusive; nonetheless, we often do not hesitate to infer meaning to their every demonstration. I think of this as I observe my dog Monty laying alongside me in a beam of sunlight streaming into my office. He looks up at me with half-slit sleep-filled eyes, and sighs. My close bond with him colors my reaction, but I think to myself how content he seems. I have no real knowledge of his internal emotional state, but his body language is revealing some cues. He is lying on the floor and his muscles appear soft and relaxed. His head is resting on the floor and the wrinkles on his forehead only appear briefly as he gazes up at me and then disappear back to their natural

Figure 4.1: The ability to read the body language of our dogs is an important skill in understanding them and safeguarding their quality of life. (Photos: R. Hack)

position. He looks at me momentarily and then looks away once I make eye contact. He closes the distance between us by extending his front paw so that it is resting, touching my foot. All of these behaviors combine to give me the overall impression that my dog is relaxed and happy. Although we like to impose our human interpretations on their behavior, there is actually now a lot of great research available on the emotions of dogs and other non-human animals,[1] and we can do more to understand how dogs really feel. We can learn to objectively read the intentions and perhaps the emotions of dogs if we carefully observe their body language.

Dogs are incredibly communicative animals, but we tend not to read them as well as they read us. Dogs are reliant on humans for their care and their expressions entice us to form indelible bonds increasing the likelihood that we care for them, love them and assure their survival. In ethology, the science of animal behavior, communication is defined as an animal's behavioral act that alters the response of another animal and is advantageous to the communicating animal's survival and fitness. In other words, dogs communicate with us to increase the chance that they will live on and that their genes are passed on to future generations. Recent studies have revealed that dogs excel in their ability to communicate and can even read human emotion. Your dog is talking to you, all the time, but are you really listening?

The science of communication

The science of dog behavior and cognition is booming and we are beginning to fully understand the interplay between dogs and humans. Work in the field of animal communication began

---

[1] Virginia Morell's fascinating book, *Animal Wise*, provides a great overview of emotion and cognition in non-human animals, from ants to rats to dolphins to chimpanzees.

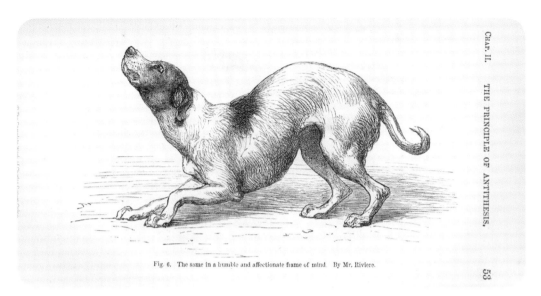

Figure 4.2: C. Darwin, illustration (1872). Darwin describes the body language of the dog pictured as "humble and affectionate." Today we would describe this as most likely an appeasement posture.

centuries ago with the first description of dogs' ability to communicate emotion in *The Expression of the Emotions in Man and Animals*. Darwin (1872) was the first to provide an authoritative account of the behavioral expression of rage, fear, and other emotions in dogs. Prior to his work on emotions in non-human animals, the dominant belief was that humans alone possessed the muscles and physical processes for demonstrating emotion. In the 60s, Scott and Fuller (1965) presented work essential to our understanding of canine genetics and emotional reactivity. They found that puppies from five distinct breeds (Terrier, Beagle, Basenji, Cocker Spaniel, and Sheltie) reared in identical conditions, had distinct differences in their reactions to stress. Terriers, Beagles, and Basenjis were consistently more reactive than Shelties or Cocker Spaniels when restrained. The more emotionally reactive breeds had higher resting heart rate levels and were more prone to express behaviors such as panting, lip licking, vocalizing, and high amplitude tail wagging. It was this evidence that led them to conclude that heredity greatly affects the expression of emotions and that differences in emotional behavior contribute prominently to the characteristic behavior of breeds. Lorenz (1981) furthered the scientific exploration of dog behavior and emotions with a fine-scaled description of canine facial expressions. He discovered that most agonistic or aggressive facial displays, short of attack, are composed of elements of both fear and aggression. Roger Abrantes later provided a great review of canine expressions and body postures, and how they can serve as social signals to maintain bonds as well as avoid damaging fights (Abrantes 1997, p. 111).

Early work in dog communication gave us a basis for understanding dog language but now we are beginning to discover dogs' unique ability to interact socially with a species very different from themselves, that is, humans. I am certain that most people that have shared their lives with a dog will not find some of the recent findings of dogs' superior social skills with humans surprising. For instance, we often think of our pet dogs as our babies. Recent studies have shown that, indeed, dogs may feel the same way about our relationship. In studies of the human–animal bond

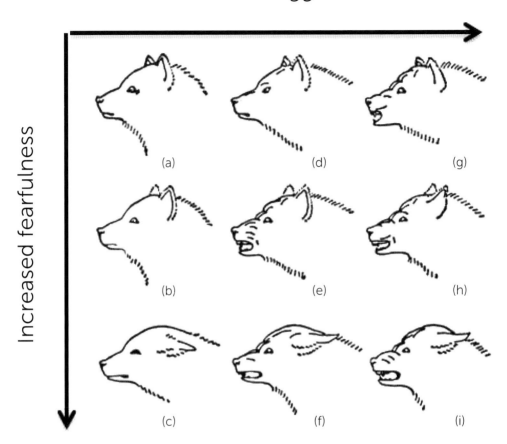

Figure 4.3: Depictions of facial expressions in dogs from increased readiness to flee (a–c) and increased aggression (a–i). (Adapted from Lorenz 1981)

in dogs, using a paradigm used to test parental attachment in children, dogs show attachment relationships to humans that resemble the bond human infants have with their adult parents (Topál et al. 1998). For instance, when dogs are separated from their human caregivers, they will play and explore less, wait by the door for their owner, will use their owners' possessions as a source of comfort (Prato Previde et al. 2003), and will enthusiastically greet their caregivers upon their return.

Not surprisingly, given this bond and the reliance dogs have on humans for their care, dogs have been shown to be very attentive to humans (Emery 2000). A number of exciting studies on dog cognition (see Chapter 3) are documenting dogs' ability to respond to cues given by humans, such as the pointing gesture or eye gaze (Gácsi et al. 2005) and even head nodding (Agnetta et al. 2000). Dogs can use social referencing (defined as the tendency to look at a significant person to gain information about an ambiguous situation). For instance, if a caregiver shows fear towards an object, dogs will inhibit their movement towards that object (Merola et al. 2012). Dogs demonstrate a strong affiliation to people, are sensitive to human behavior and adjust their behavior based on

Relaxed facial expression (note open mouth, relaxed ears, soft eyes)

 Relaxed/playful (note open mouths, relaxed stance, soft eyes)

 Playful (note play bow – an invitation to play and/or signal that the next moves are meant as play)

 Alert, interested, somewhat unsure (note closed mouth, upright ears, focused gaze, slight head tilt)

 Nervous (note lowered posture, shrinking away, lowered ears, closed mouth, 'whale eye', 'worried' brow)

 Fearful (note lowered posture, mouth closed, 'whale eye', head turned away, lowered ears)

 Defensive aggression (note lowered posture, hard stare, lowered ears, teeth bared, lip lifted)

 Aggressive/warning (note 'agonistic pucker' – teeth bared and lip lifted, hard stare, ears up and forward)

Figure 4.4: Canine communication: facial expressions and body postures. (Photos: iStock/Getty Images)

Figure 4.5: Domestic dogs have evolved to be very attentive to humans; this is particularly clear when watching trainers work with their dogs. (Photo: K.D. Setiabudi)

Reading your dog 55

human reactions (Duranton and Gaunet 2015). Interestingly, a number of studies have shown that despite dogs' ability to read human communication in order to accomplish a goal (such as locating hidden food), dogs have great difficulty in understanding humans' intentions unless these intentions are clearly communicated in a way that dogs can understand. In studies in which dogs must interpret the goal of humans in order to assist them, dogs cannot perform such tasks without the assistance of ostensive cues (Kaminski et al. 2009; 2012). Humans also have difficulty in reading intentionality in dogs, a finding supported by a number of studies on the risk factors for dog bites. This inability to read our dogs can, and often does, lead to unfortunate outcomes in dogs such as aggression and even biting; these events in turn can lead to a breaking of the dog–human bond that puts pets at risk of rehoming or euthanasia. If we wish to provide dogs with the most humane treatment, we must learn to listen to what they are saying, to observe our dogs and learn their language so that we can both assure their well-being and avoid costly mistakes. Knowing when our dogs are uncomfortable is also an important baseline skill for successfully training and socializing our dogs; more on this in Chapter 5.

Reading and responding to your dog

Learning to read your dog begins with careful observation. If we were as good at reading them as they are at reading our signals, we would have much more efficient communication and avoid mistakes. Many dog owners do not recognize the subtle signs of anxiety, arousal, fear, or distress in their dogs. By learning how dogs communicate these emotional states we can respond more humanely, reduce risks of fear-based aggression, and build a stronger bond with our canine companions.

Dogs use every part of their body to signal their internal emotional state, so let us begin by breaking down signals by body part (Table 4.1).

Eyes

A dog's eyes can relay information through both the size and shape of the eye or in the duration, target, and intensity of eye contact.[2] Although the shape of the eye conveys information about the emotional state of the dog, the direction and duration of the gaze may alter the meaning of the visual display.

- Shape:
  - Happy: A relaxed, content dog will have wide-open eyes with soft direct or averted gaze. The skin around the eyes will be relaxed and show no wrinkling. Some dogs' eyes are round while others appear more oval. This is what we consider to be the normal shape of the eye.
  - Fearful: A fearful dog may have partially closed or squinting eyes. An insecure, fearful, or appeasing dog's eyes will become smaller and more elongated. In a fearful dog you

---

[2] The size and appearance of the dog's pupil is another indicator of the dog's internal state. Dogs that are showing arousal will have dilated pupils.

Table 4.1 Emotional state signalled by body part. (Illustrations by H. Light).

Body part	Happy	Fearful	Aggressive
Eyes	Soft with direct eye contact but will avert	Averted, squinting or overly wide	Tense, open wide, the direct contact or hard stare
Mouth	Mouth gently closed or slightly open, relaxed	Lips pulled back "grin," licking lips, yawning	Lips puckered in front, lips lifted to reveal teeth, bared teeth
Ears	Forward or slightly back, neutral position	Low or pinned back	Forward or pinned back in threat
Tail	Low and still or medium carriage, wagging	Tucked under or close in, fast wag	High carriage, flagged, still or slow wag
Hair	Soft, flat	Excessive shedding, piloerection	Piloerection

may see wrinkling of the skin around the eye or on the face. A dog that is ill or in pain will often close his eyes or squint. A squinting dog may also be seen in a happy greeting so as with all body language, you need to study your dog, consider the context, and look for other cues to understand his meaning. Not all fearful dogs have partially closed eyes; a very fearful or conflicted dog may also have eyes that appear larger than normal with whites of the eye visible. This eye merits caution.

- Aggression: A dog that is displaying aggression or that is very aroused has eyes that are open wider than normal. You may notice more white around his eye because of this and his pupils may be dilated (appearing larger than normal). The skin around and between the eyes is tense. Wrinkling of the forehead is common. Eyebrows appear to be pulled inward making them seem more prominent. This eye also merits caution.

- Duration, target and intensity:
  - Happy: A dog may look directly at you with relaxed, open eyes in an attempt to elicit your attention. A direct stare has often been described as threatening behavior, yet most dogs learn that it is okay to look directly at people, as a result of this behavior having been rewarded in some way in the past (such as by petting, verbal praise, or even food). A direct stare with relaxed face and eyes, often open mouth, can signal that the dog is being affiliative. Dogs that are happy will look away or avert their eyes if you make prolonged eye contact. Their gazes are shorter, softer, and less direct than a dog that is fearful or aggressive.[3]

---

3 In a study of behavior testing at a shelter, dogs that looked away from the tester during a direct stare test were found to be less likely to show aggression after adoption (Donaldson 2010).

Figure 4.6: Two dogs with very different opinions about their party hats. (Photo: R. Hu)

- Fearful: A dog that averts his gaze when faced with a direct stare is signaling that he is willing to conciliate. It can also indicate that he is worried about interacting with you. Avoiding eye contact may signal fear or a lack of confidence or experience. Conversely, a fearful dog may also maintain direct eye contact with the object of their anxiety while physically positioning their head or body away, resulting in a form of eye contact called "whale eye"), where the sclera (or whites of the dog's eye) are visible. An extremely fearful dog may show eye contact similar to an aggressive dog if the fear is escalating to fear aggression.
- Aggressive: Aggression, in a behavioral context, has a number of different definitions and functions. In the animal world, and with the exception of predatory aggression, that goal of aggressive behavior is often to increase the distance between the aggressor and a perceived threat. An aggressive dog often displays direct, prolonged eye contact with the person or object of threat. In these dogs, the whites of the eyes are visible and the pupils are often dilated. Eye contact is sustained; the dog may not avert his eyes from his target even though he may turn his head; whale eye may be seen. If your dog turns his head away but moves his eyes toward the perceived threat, so that the eyes are open wide, often with whale eye, he may be preparing for an attack. Stiff and rigid body muscles, raised hackles, and growling, often accompany this eye contact. Dogs showing this body language are asking for space and that is exactly how you should respond! When you encounter this signaling, look away and create distance between you and the dog by slowly backing away in order to reduce the threat that the dog perceives. These dogs can be very unpredictable and possibly dangerous (particularly if you ignore their request for more space).

Mouth

Although dogs lack the ability to speak human language, they do convey information with their mouths. From the shape, position, and tension of the mouth, to vocalizations, the mouth can reveal a lot about a dog's emotional state.

Figure 4.7: Relaxed facial expression. (Photo: K.D. Setiabudi)

- Shape: Dogs position their lips and jaws and display their teeth in a variety of ways to communicate. Dogs also convey information by various sounds, as described below.[4]
    - Happy: A relaxed, content dog will have a mouth that is closed or slightly open. The muscles around the muzzle will be loose and soft. There will be no pulling back at the corners of the mouth (called the commissure) and there will be no puckering at the front of the mouth. Either the lines around the mouth are not visible or there will be a few that appear soft when the mouth is open and the dog is panting.[5] A happy dog may produce a short repeated whine or a quick, high pitched bark to invite attention or play.
    - Fearful: A dog that is fearful or overtly submissive will have a closed mouth, with muscles around the mouth tense, or will have just the corners of his mouth slightly pulled back in a submissive grin. A submissive grin is defined as a pulling of the lips vertically exposing only the front teeth; the appearance is that of a smile but the motivation is not one of pleasure, as many people assume. This expression may be combined with a lowering of the head and/or body appearing to cower. This display is intended to dispel a perceived threat (what is sometimes called an appeasement gesture). An unknowing person may mistake this for an aggressive display, or conversely, as an invitation for attention, despite the fact that this is the opposite of what the dog is attempting to convey. Dogs displaying signs of fear are likely to react with a threat behavior if the fearful display does not work to remove the perceived threat. A dog displaying these signs is asking for distance, so give this dog space. A fearful dog may also whine, but the whine is longer

---

4   See Abrantes 1997. *Dog Language: An encyclopedia of canine behavior* pp. 231–232, for a complete description of canine vocalizations.
5   Dogs pant to cool off and maintain a comfortable body temperature but they also increase panting when stressed or aroused. This form of panting is different than that of a stressed dog. Stress panting is more intense and is accompanied by a tighter muscled mouth with greater wrinkling at the commissure.

and rises in pitch, bark, or growl to communicate his need for space. Fearful or stressed dogs also yawn. Yawning in dogs is sometimes seen as a signal that dogs use to pacify a threatening dog but it may also signal a general sense of anxiety. If your dog is yawning, sometimes accompanied by a whine, he may need space and/or time to relax.

- Aggressive: A dog signaling a threat of aggression will fully retract his lips vertically while at the same time wrinkling his muzzle. Aggressive dogs will also retract their lips both horizontally and vertically so that the front teeth as well as the back will be on display (for how better to intimidate a potential rival or threat than by showing them all your teeth!). This too is a warning, telling the person or dog at the receiving end of this signal to stay away. Another display is the "aggressive pucker." Dogs may move their lips forward over their teeth and exhale air so that their lips look puckered and large. There is often audible breathing and tight facial muscles with increased wrinkling on the forehead or between the brows. This look also means stay away, and dogs resorting to this are likely to bite if pushed any further. Aggressive dogs impart threat through vocalizations as well. The aggressive growl is a deep, low tone growl. A similar vocalization with raised lips is the snarl. The bark of an aggressive dog is quick, deep, and repeated often until the object of the threat retreats or the dog escalates to a bite.

It is worth remembering at this point that one very important function of aggression is communication: by paying attention to dogs' signals and respecting them, we can better understand how our dogs perceive the situation, and reduce risk of bites or damage that can have fatal consequences, for the dogs themselves as well as any human or other targets of that aggression. Our primary goal in any aggressive interaction with a dog should always be to get everyone involved out of the situation safely; we'll talk about this more in the next section of this chapter, on staying safe around dogs.

Figure 4.8: Same dog, different day ... can you interpret the underlying emotional state of this dog in the two images, based on body language? (Photos: S. Watko)

Figure 4.9: Ear position, and other facial clues, can tell us a lot about a dog's current emotional state. Note in particular the dog on the right: ears back, mouth closed, and commissure back, and looking away from the other dog. (Photo: K.D. Setiabudi)

Dog ears

Dogs have a wide variety of ear types from hanging to pricked and everything in between. Cropping the ears to remove some or all of the earflap can also alter communication with the ears. Given the variety of dog ears, careful observation of your dog is necessary to fully understand their communication. The position of the ear can reveal some clues to emotional state.

- Position:
  - Happy: A relaxed, content dog will hold his ears in a natural position. When he is alert, he will raise them higher on his head and he will direct them towards whatever is holding his interest. If your dog has his ears pulled back slightly, he's signaling his intention to be friendly. Dog ears that are alternating position may signal that he is attending to the world around him.
  - Fearful: A fearful dog will generally hold his ears back, flat against his head.
  - Aggressive: An aggressive dog will hold his ears erect and forward when feeling assertive.

Dog tails

Dog tails represent the bedrock of communication to most people. We learn at a very young age that dogs wag their tails when they are happy. Although this is true, dogs also wag their tails in a variety of other contexts, including aggression. Recent studies even show that the direction of

Reading your dog 61

Figure 4.10: Two dogs meeting at a doggy day-care facility; what can you discern about the emotional states of the two dogs in each photo, based on body language? (Photos: K.D. Setiabudi)

the tail wag reveals information about the emotional state (Quaranta et al. 2007). Researchers found that a tail wag that is primarily to the (dog's) right signals that the dog is happy that you are approaching, but a tail that is wagging primarily to the left may signal fear or apprehension at the approach of a person. In addition to the directionality of the wag, dogs' communication with the tail depends on their tail carriage. Most breeds of dog have a natural tail that extends outward or toward the ground. Some dogs hold their tails more curled and up (think Akita) and some hold them down and possibly tucked (think Greyhound). Some breeds such as the pug have tails curled close to the body. Some dogs have naturally bobbed tails, such as Australian Shepherds, and some have surgically docked tails, such as the Boxer. As with the ears, careful observation of your own dog in a variety of situations will be required to determine what he is communicating with his tail.

- Tail carriage:
  - Happy: A relaxed, content dog will hold his tail in a natural carriage or will wag his tail gently from side to side. If the dog is excited he may increase the force of the wag, the speed of the wag, or move the tail in a circular pattern.
  - Fearful: A dog that is fearful will often hold his tail low and still, or tuck the tail close to the body or between the legs. Many times, fearful dogs wag their tails but at a more rapid pace than that of a relaxed dog. A dog that is alert or aroused often holds his tail higher than normal. Dogs will hold their tails with a high carriage or "flag" if they are alerting to a movement of noise. There will be tension in the tail and it may appear to vibrate.

- Aggressive: An aggressive dog may stop all movement of his tail while standing still. This dog may also hold their tail in a high carriage and "flag." Dogs may also wag their tails stiffly, or at a faster than normal pace while responding to a threat.

Fur or hair

Some clues to the emotional state of the dog are revealed by the condition of the fur or hair. There are a couple of ways that a dog's emotional state may be communicated by the coat. Dogs that are stressed will shed excessively or "blow their coats"; you may notice that your dog sheds more at the veterinarian's office or his first time in a new place. This is an indicator that your dog is stressed and may need some time or space to relax.

Another way a dog may communicate with its coat is through piloerection. During piloerection, the hairs are generally raised around the shoulders, neck, or withers of a dog. This is often referred to as "raising the hackles," and it can be an autonomic response to fear, aggression, anxiety, excitement, or general arousal. If you notice that your dog has raised hackles, give your dog space or time to relax.

Body posture

You can tell a lot about the emotional state of your dog through its body posture.

- Happy: A relaxed, content dog will usually stand at his full height with his head held in its natural position. A happy dog will have loose, fluid movement and relaxed muscle tone.
    - A playful dog will have exaggerated body postures and movements. He may be bouncing around or running inefficiently or may be in a submissive posture, but his facial expression and his muscles will be relaxed and nothing about his body will look unnatural. A playful dog may give high-pitched short barks in an attempt to engage you in play.
    - Fearful: A fearful dog will attempt to make his body appear smaller to avoid threat. This dog will lower itself, cower, or appear to cringe. The back may be curved with the head

Figure 4.11: Play bow. (Illustration: H. Light)

Figure 4.12: Fearful posture. (Illustration: H. Light)

slung low. The tail is carried low or tucked. The ears will be held down and back and the face will be averted. The ears will be held down and back. The muscles of the body will often appear tense or stiff.

- ◆ Aggressive: An aggressive dog will attempt to make his body appear as large as possible. He will stand erect at full height, sometimes even on his tiptoes. He will hold his head high, with his neck raised above his shoulders. His weight will either be centered over all four feet or he will be leaning slightly forward on his front legs. His tail will be erect and high. The loose, fluid movements of a happy, relaxed dog are not in evidence; an aggressive dog will be tense and focused, often with a stiff body posture.

Figure 4.13: Confident/assertive posture. (Illustration: H. Light)

How to interpret emotional states of your dog to keep yourself safe around dogs

Dogs communicate important information that is often overlooked. As their caregivers, we need to be committed to learning to recognize how the dog is feeling and how to respond in a way that is both appropriate to the context and humane to the dog. By becoming proficient in dog body language and recognizing emotional states in your dog, you will not only build a more trusting relationship but also avoid unintended conflict that can lead to aggression. In his book, *Outliers: The story of success*, Malcolm Gladwell asserts that it takes around 10,000 hours to become an expert (Gladwell 2008). Even if you have natural talent, without practice you will never achieve greatness. The elite do not simply work at something; they fall in love with practicing that skill. Apply this to your education with your dog. Fall in love with reading your dog. Note what makes him show signs of anxiety, and what brings him calm and contentment.

Happy

How do we know when our dog is happy? In Chapter 2, we discussed some ways of evaluating well-being in dogs, but here we will discuss what you can do to determine through your dog's body language if your dog is happy. When your dog is happy, he gives an overall appearance of relaxation. He may show various levels of positive arousal or excitement, dependent on the context. You are likely to be very familiar with the excited, happy dog you see when you arrive home from a long workday away, or if your dog is like my dog, perhaps just a few minutes away. Overall, his muscles are relaxed, his movements are fluid, his tail and ears are held in their natural positions or forward in greeting. A happy dog will wag his tail from side to side or in a circular motion. His facial expression is neutral, the muscles in his face are relaxed, and mouth is closed or slightly opened. The corners of his mouth (commissure) might be turned upwards slightly, as though he's smiling.

Calm and relaxed

To our benefit, most pet dogs are in a calm and relaxed state most of the time. A calm dog will lie or sit quietly in your company. Calm dogs have slow, measured respirations, soft muscle tone, neutral ear and positions, and relaxed facial muscles. Dogs may smack their lips or gently blink their eyes.

Excited and playful

When your dog is excited, he looks as intense as he does when he's alert, but he might also adopt a playful demeanor. His body is ready for action. His weight might be centered over his rear legs as he prepares to move. His ears are up and his tail is held high, and it may or may not wag. Excited dogs often hold their mouths open, and they might bark. So, how do you know when your dog is playing or quite serious? Dogs retain many playful behaviors well into adulthood or possibly for the duration of their lives (Bekoff 1974; 1977; 1995). Dogs that are playful may take on postures of a confident dog such as a high tail carriage, direct eye contact, and even play barking. However, playful dogs solicit play through movements that are exaggerated and seem to put the dog at a disadvantageous position. Dogs enjoy a variety of play styles, including chase games (in which the dog is either the chaser or the chasee), rough-and-tumble (wrestling or tackle) games, and games of "keep-away" with an object, like a toy or stick. Almost all play is interspersed with the characteristic "play bow" that's common across all dogs. When your dog play bows, he bounces into position with his forelegs on the ground and his hind legs extended so that his rear end sticks up (like the pose known appropriately as the "downward dog" in yoga practice). This signal is extremely important because so much of dog play consists of aggressive behaviors and dominant postures. The play bow tells a dog's playmate, "Anything that comes after this is play, so please don't take it seriously." Some dogs also show a "play face," a happy facial expression characterized by a partially open mouth that almost looks as though the dog is smiling. A playful dog might also growl or make

Figure 4.14: Georgia has spotted something in the tree, and is on alert. (Photo: M. Marchitello)

high-pitched barks but the meaning of these vocalizations is altered based on the playful context. Dogs may also play bite (place teeth with little or no force and open jaw or mouth you during play). These behaviors should be closely monitored because, even in play, dogs' teeth are sharp and can cause damage to human skin. Caution should also be exercised during play so that play aggression does not tip over into true aggression if arousal is too high. When playing with your dog be careful to watch your dog for signs of increased arousal (harder play bites, stiff body, prolonged growling, or barking). If your dog becomes overexcited, take a break and let your dog relax.

Alert

When your dog is alert, he looks intense and focused. He stands upright with his weight centered on all four feet, his ears are up and forward, and his head and neck are erect. He holds his tail either in its natural position or vertically, possibly even over his back. His tail is rigid and immobile. He may have piloerection (raised hackles). His eyes will be focused and he may not divert his attention to you. His mouth is typically closed. He may produce a growl, bark, or whine. An alert dog may be confident, fearful, or aggressive, so be watchful for signs of further arousal.

Aroused

When your dog is aroused, you might have a hard time distinguishing it from when he is simply excited. It is important to notice signs of arousal because increased arousal can lead to reactions based on fear or aggression. Some signs to monitor are increases in piloerection, decreases or increases in breathing, increases or decreases in movement, inattention to commands, staring and freezing (complete cessation of movement). The loose, relaxed body posture of a happy, excited dog is not seen; instead, muscles are tense and ready for action. These are signs that your dog may tip over into aggression or flee in fear. If you notice signs of arousal, consider the context that the dog is in, and if necessary, create distance for your dog by moving in the opposite direction from the target of his arousal. Attempt to get the dog's attention; if this is not possible, move the dog further from the target of arousal. Work to calm the dog by asking for behaviors that encourage relaxation, such as a sit or a down and reward your dog for calm behavior. When at home or in a situation in which your dog is relaxed, practice having your dog make eye contact with you *on cue* so that you gain control over his eye contact for future encounters. Practice relaxation postures[6] so that your dog is more proficient at calming down from an aroused state. Seek professional assistance if the arousal continues to escalate or you observe signs of fear or aggression.

Anxious or fearful

As caregivers to our dogs, we sometimes fail to recognize the danger of anxiety or fear in our dogs. Fearful dogs or ambivalent dogs are unpredictable. A fearful dog may be more likely to deliver a bite than a more confident, aggressive dog. If your dog is feeling anxious or fearful he

---

6 Such as sit, (lie) down, go to your mat/bed, etc.

may move more slowly and lower to the ground. Fearful dogs try to avoid sources of anxiety, so you may notice that he turns his head away from what scares him, or positions his body to avoid the perceived threat. Dogs often avert their eyes or turn their head away while still keeping an eye on the source of anxiety, resulting in the "whale eye" look described earlier. They may sit or lie down, exposing their underside or other vulnerable parts; this is a submissive, appeasement posture designed to "turn off" aggression in an approaching threat. They may lip lick, nose lick, show excessive eye blinking, chatter their teeth, drool, pant, sweat from paws, tremble, whine, yawn, or chew on or groom their bodies or paws. These are called displacement behaviors and we see them in dogs with high anxiety. Never force your dog to confront the thing that scares them; this is a very common mistake that many well-intentioned dog owners make, that runs a high risk of backfiring (making the dog more fearful) and can be very stressful for the dog (not to mention risky if the dog is so fearful that he becomes aggressive). Give them space, and when time allows, work slowly to desensitize and countercondition[7] them to improve their confidence. Far too often I encounter well-meaning dog owners that say their dog loves children and invite a child to approach. What I see is a dog that is doing its best to avoid the child, moving in the opposite direction, hiding by the owner's leg, crouching to the ground. This is a recipe for a bite. Just because your dog tolerates something, does not mean he likes it. Watch and listen to your dog. He may not share your opinions about what is safe and what is not. We cannot solve the problem by simply telling our dogs that they don't need to be afraid,[8] any more than we can cure a person of fear of flying by simply telling them to "get over it." Work with a certified applied animal behaviorist (CAAB), veterinary behaviorist or a certified pet dog trainer[9] to develop a training protocol to decrease your dog's fear and develop confidence.

Aggressive

A dog's threat displays comprise an array of body postures, facial expressions, and vocalizations. Identifying these signals can mean the difference between a bite and a safe interaction. Aggressive postures include an upright forward orientation with muscles tense and still, head up or horizontal, ears upright and immobile or flattened, eyes staring or fixed, piloerection, mouth open and taut, teeth exposed, prominent wrinkling of the forehead, growling, and barking (Abrantes 1997; Aloff 2005; Simpson 1997). This behavior is a normal component of dog behavior but can be dangerous if practiced in the wrong context. Signs of aggression vary depending on the level of arousal and fear. Stressed dogs may show heightened responses such as more raised hair and growling than non-stressed dogs (Beerda et al. 1999). The British Small Animal Veterinary Association (BSAVA) has developed a "ladder of aggression" to assist dog owners in recognizing the steps in how dogs respond to a threat (see Figure 4.15). The goal should always be to keep your dog from moving *up* the rungs of the ladder; a calm, relaxed dog is a much safer dog.

---

7   We will talk more about the process of desensitization and counterconditioning in Chapters 5 and 6, on training and mental wellness.
8   Although we can certainly help the situation, or make it worse, by our own behavior and body language in that situation; more on this in Chapter 5.
9   For more on sources of expert help, and for examples of who we consider "qualified," see Chapter 7, on behavior problems and solutions.

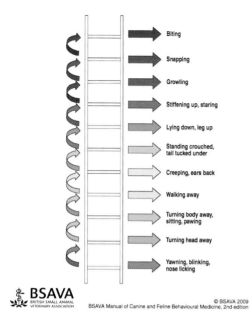

Figure 4.15: The canine ladder of aggression (K. Shepherd): How a dog responds to stress or threat. Source: Horwitz, D. and Mills, D. (2012) *BSAVA Manual of Canine and Feline Behavioural Medicine*. Gloucester, UK: British Small Animal Veterinary Association. Reproduced with permission from the *BSAVA Manual of Canine and Feline Behavioural Medicine*, 2nd edition. © BSAVA

Fear and aggression are closely linked in dogs; for some dogs, if their attempts to diffuse a threat through communication fail, they may feel forced to resort to defensive aggression. The problem for many dogs is, once they learn that this aggression is an effective way to get the threat to move away, they will continue to use this strategy in the future.

Dogs may show some ambivalence in their body language and still be capable of attack. Fear and threat may both be present in the same facial expression or body posture. An ambivalent dog may have lips drawn back horizontally and taut at the commissure. This is primarily a look of fear or uncertainty. A dog showing this facial expression is also signaling the desire for you, or whatever the threat may be at that moment, to stay away, but there is a lack of confidence. Dogs that move quickly forward and then retreat, or change from one body posture to the next, are showing uncertainty. Dogs are often reported to be wagging their tails preceding an attack; in this scenario, tail wagging is a signal of arousal rather than greeting. A fearful dog may cower with weight shifted back and ears in a low position prior to attack (Aloff 2005; Quaranta et al. 2007). Vocalizations also vary depending upon the emotional state evoked in different contexts (Faragó et al. 2010). For instance, a dog that is showing appeasement or that is soliciting caregiving may produce a high pitched whine; whereas a dog that is displaying aggression may elicit a low pitched growl. Dogs usually exercise restraint in aggression. They mostly rely on threats such as a growl and avoidance rather than escalating to biting attacks in most cases. Luckily for us, dogs learn to inhibit aggression through early experiences within the litter (Fox 1970) and thwart aggressive attacks through signalling of or deference. However, if you observe these signs in your dog, remove your dog to avoid the situation until you can seek professional help of a CAAB, veterinary behaviorist, or certified pet dog trainer that has been trained to treat behavior issues in order to desensitize and countercondition your dog.

Fearfully aggressive

If your dog is fearfully aggressive[10] he might not look much different than when he's fearful, except that he might show increased levels of arousal. He might show his teeth and growl. Some

---

10 Fearfully aggressive dogs are defined as dogs that show aggressive behavior toward unfamiliar humans. This behavior may also be termed as stranger aggression or aggression to unfamiliar people.

fearful dogs never escalate to aggression, but others will if they have displayed threats and the object of their fear does not retreat. A fearful dog may continue to stay low to the ground, ears back, eyes averted or with a "whale eye." He may display his teeth and produce a snarl or a growl. Fearfully aggressive dogs will often snap (intentional bite in the air) or bite. Some fearful dogs will bite the object of their fear from behind once it retreats. Observe your dogs for signs of fear and be respectful of what your dog is trying to communicate. Respond to his signals and give the fearful dog distance before his arousal increases and he becomes fear aggressive. Fear aggression is a serious problem in dogs and should be carefully treated with the help of one of the trained professions mentioned above.

Take-home messages

1. Dogs are excellent readers of our body language; and we can observe and learn to read them as well. Being able to accurately "read" a dog is an essential skill for anyone living and working with dogs.
2. Dogs perceive the world differently than we do. They do an incredible job reading us, responding to us and living amongst us, but we must be careful not to misread their communication by superimposing our own emotional reactions on their behavior.
3. Learn to observe signs of stress, fear, and anxiety in your dog and respond humanely. The "ladder of aggression" is a great tool to help visualize and understand the progression from mild fear to serious aggression in dogs faced with stressful situations. The goal should always be to help the dog relax, and prevent the dog from moving up the "ladder."
4. If your dog is showing signs of significant fear, or aggression, get help as soon as possible from a trained, certified professional in behavior modification. Additional sources of background information on understanding and treating serious behavior problems can be found in the "Annotated bibliography" section at the end of this book.

References and additional resources

Abrantes, R. (1997). *Dog Language: An encyclopedia of canine behaviour*. Naperville, IL: Wakan Tanka Publishers. 266 p.
Agnetta, B., Hare, B., & Tomasello, M. (2000). Cues to food location that domestic dogs (Canis familiaris) of different ages do and do not use. *Animal Cognition* 3: 107–112.
Aloff, B. (2005). *Canine Body Language: A photographic guide*. Wenatchee, WA: Dogwise Publishing. pp 110–111.
Beerda, B., Schilder, M.R.H., & van Hoof, J.A.R.A.M. et al. (1999). Chronic stress in dogs subjected to social and spatial restriction. I. Behavioral responses. *Physiology and Behavior* 66: 233–242.
Bekoff, M. (1974). Social play and play-soliciting by infant canids. *American Zoologist* 14: 323–340.
Bekoff, M. (1977). Social communication in canids: Evidence for the evolution of a stereotyped mammalian display. *Science* 197: 1097–1099.
Bekoff, M. (1995). Play signals as punctuation: The structure of social play in canids. *Behavior* 132: 420–429.
Darwin, C.R. (1872). The expression of the emotions in man and animals. London: John Murray. p 53.
Diesel, G., Bordbelt, D., & Pfeiffer, D. (2008). Reliability of assessment of dogs' behavioural responses by staff working at a welfare charity in the UK. *Applied Animal Behaviour Science* 115: 171–181.

Donaldson, T.M. (2010). Behavioral assessment of aggression towards humans in the domestic dog. Washington State University.

Duranton, C., & Gaunet, F. (2015). Canis sensitivus: Affiliation and dogs' sensitivity to others' behavior as the basis for synchronization with humans. *Journal of Veterinary Behavior: Clinical Applications and Research* 10: 513–524.

Emery, N.J. (2000). The eyes have it: The neuroethology, function and evolution of social gaze. *Neuroscience and Biobehavioural Reviews* 24: 581–604.

Faragó, T., Pongrácz, P., Range, F., et al. (2010). 'The bone is mine': Affective and referential aspects of dog growls. *Animal Behaviour* 79: 917–925.

Fox, M.W. (1970). Overview and critique of stages and periods of canine development. *Developmental Psychobiology* 4: 37–54.

Gácsi, M., Gyori, B., & Miklósi, A., et al. (2005). Species-specific differences and similarities in the behavior of hand-raised dog and wolf pups in social situations with humans. *Developmental Psychobiology* 47: 111–122.

Gladwell, M. (2008). Outliers: The story of success. New York, NY: Little, Brown and Co. 309 p.

Hare, B., & Tomasello, M. (2005). Human-like social skills in dogs? *Trends in Cognitive Sciences* 9: 439–444.

Hare, B., Brown, M., Williamson, C., & Tomasello, M. (2002). The domestication of social cognition in dogs. *Science* 298: 1634–1636.

Kaminski, J., & Marshall-Pescini, S. (2014). *The Social Dog: Behavior and cognition*, 1st Edition. London: Academic Press.

Kaminski, J., Templemann, S., Call, J., & Tomasello, M. (2009). Domestic dogs comprehend human communication with iconic signs. *Developmental Science* 12: 831–837.

Kaminski, J., Schultz, L., & Tomasello, M. (2012). How dogs know when communication is intended for them. *Developmental Science* 15: 222–232.

Kerepesi, A., Jonsson, G.K., Miklósi, Á., et al. (2005). Detection of temporal patterns in dog-human interaction. *Behavioural Processes* 70: 69–79.

Lockwood, R. (1995). The ethology and epidemiology of canine aggression. In J. Serpell (Ed.), *The Domestic Dog: Its evolution, behaviour and interactions with people*. Cambridge: Cambridge University Press. pp 131–138.

Lorenz, K. (1981). *The Foundations of Ethology*. Chapter VII. Multiple motivation in behavior. New York: Springer Science and Business Media. p 244.

Merola, I., Prato-Previde, E., & Marshall-Pescini, P. (2012). Dogs' social referencing towards owners and strangers. *PloS One* 7: e47653.

Miklósi, Á. (2007). *Dog Behavior, Evolution, and Cognition*. Oxford: Oxford University Press.

Miklósi, Á., & Soproni, K. (2006). A comparative analysis of animals' understanding of the human pointing gesture. *Animal Cognition* 9: 81–93.

Miklósi, Á., Kubinyi, E., Topál, J., et al. (2003). A simple reason for a big difference: Wolves do not look back at humans, but dogs do. *Current Biology* 13: 763–766.

Morell, V. (2013). *Animal Wise: How we know animals think and feel*. New York, NY: Broadway Books. 291 p.

Prato Previde, E., Custance, D.M., Speizio, C., Sabatini, F. (2003). Is the dog-human relationship an attachment bond? (An observational study using Ainsworth's strange situation). *Behaviour* 40: 225–254.

Quaranta, A., Siniscalchi, M., & Vallortigara, G. (2007). Asymmetric tail-wagging responses by dogs to different emotive stimuli. *Current Biology* 17: R199–R201.

Reid, P.J. (2009). Adapting to the human world: Dogs' responsiveness to our social cues. *Behavioural Processes*. 80: 325–333.

Scott, J.P, & Fuller, J.L. (1965). *Genetics and the Social Behavior of the Dog*. Chicago: University of Chicago Press.

Simpson, B.S. (1997). Canine communication. *Veterinary Clinics of North America* 27: 445–464.

Soproni, K., Miklósi, Á., Topál, J., & Csányi, V. (2001). Comprehension of human communicative signs in pet dogs (Canis familiaris). *Journal of Comparative Psychology* 115: 122–126.

Topál, J., Miklósi, Á., Csányi, V., & Dóka, A. (1998). Attachment behavior in dogs (canis familiaris): A new application of Ainsworth's (1969) Strange Situation Test. *Journal of Comparative Psychology* 112: 219–229.

Chapter 5

The science of humane dog training

Photo: iStock

WE AS DOG OWNERS AND DOG ENTHUSIASTS ARE VERY FORTUNATE TODAY, AS there is a wealth of solid, fact-based and research-supported information out there describing why dogs behave in the ways that they do, and how animals, including dogs, learn. The principles of learning in animals (including humans) have been well-studied and well-documented, and modern-day professional dog trainers and canine behaviorists make daily use of this knowledge. Familiarizing ourselves with these basic principles helps us understand *how* dogs learn, which in turn helps us to get them to learn what we want them to learn.

Taking these principles into account when working with our dogs inevitably helps our training be more effective and successful, and can help us predict how a dog will behave in a given situation, faced with a certain trigger or given set of circumstances. Knowledge of these principles can also help us understand what happened when our efforts at training fail: why did our dog not do what we hoped he would do? In this chapter, we will cover the basic principles of learning that are most applicable to training dogs: habituation and associative learning (classical conditioning and operant conditioning), as well as a few other useful tools such as shaping and capturing. For a more in-depth look at how dogs learn, particularly operant conditioning and the mechanics of training new behaviors, we recommend the appropriately titled book *How Dogs Learn*, by Mary Burch and Jon Bailey.[1]

Be familiar with the scientific principles of how dogs learn; these are great tools for maintaining a calm and happy home with your dog

The first type of learning we'll cover is habituation, a very simple form of learning. Living organisms are built to respond to stimuli in their environment, to the sounds, sights, smells of the world around them. Sometimes these stimuli alert us to good things, like food or friends; other times, they alert us to potential danger, like an approaching threat or a harmful substance. This ability to recognize and respond to stimuli is a good thing; it allows us to survive and prosper in our world. However, not all stimuli are equally important to us; some events and objects in our environment are effectively meaningless for us – they do not influence our lives in any significant way. For this reason, most organisms have evolved ways to filter out unnecessary information, so that we can focus on information that we need, and avoid wasting precious energy in constantly reacting to harmless stimuli. Habituation is one such filtering method, and is defined as the process by which, after a period of repeated or continual exposure to a stimulus, an animal's response to that stimulus decreases or stops.

Our dogs do this, and so do we. As an example of this process in action, when I was a graduate student in San Francisco, California, I moved into an apartment right next to a busy fire station. Periodically, a speaker in the station would suddenly (and loudly) squawk out details of a fire or accident in progress, calling the firefighters into immediate action. Initially, this loud noise never failed to startle me; but never (fortunately for me) did the announcement apply to me or anyone I knew. After a while, I stopped noticing the announcements; it became background noise. Visiting friends would startle visibly and ask, "What is that?!" and it would take a moment for me to register that an announcement was in progress. I had habituated to the stimulus of the fire station loudspeaker; my brain was able to tune out the noise, and I simply didn't hear it anymore. Habituation is fairly stimulus-specific, however. Just because I was able to tune out the sudden, loud sound of the fire station loudspeaker, it did not (unfortunately) mean that my neighbor's car alarm blaring at midnight went unnoticed! This goes for our dogs as well; just because they no longer react to the familiar sound of the neighbor's dog barking, does not mean

---

1 Burch, M.R. and Bailey, J.S. 1999. *How Dogs Learn*. Howell Book House.

they will never react to the sound of an unfamiliar dog barking when they are out on a walk, or to a cat fight underway in the neighbor's yard, and so forth.

Practical applications of habituation: why Puppy Class is a great idea

There are certainly times that we want our dogs to habituate to a certain stimulus, and we can help our dogs to habituate to many commonplace stimuli by ensuring that they are introduced to them in a positive, non-threatening way regularly, frequently, and as early in their lives as feasible. In general, dogs are quicker to habituate when they are very young (Hart et al. 2006). It is for this reason that well-run puppy classes, in which young dogs are introduced to many new people, dogs, objects, and situations in a positive, non-threatening way, are so helpful in shaping calm adult dogs. As we will talk about more in Chapter 6, fear and aggression are often linked in dogs, and for dogs, unfamiliar things are often scary. As a general rule, the more things in the world that are made familiar (and non-threatening) to a young dog, the fewer things that will make that dog fearful, and the calmer that dog is likely to be later in life. This is why socializing your new puppy is essential; see Boxes 5.1, 5.2 and 5.3 for some tips on this process, and Box 5.4 for information on when to socialize. In addition, it is worth bearing in mind that some stimuli

Box 5.1. Socialize, socialize, socialize! Some tips for socializing your puppy

1. When to socialize: As your puppy develops, the important window of her life during which you should actively socialize her is between 3 weeks and 3 months of age (up to 16 weeks). Before your puppy meets any new dogs or goes out into the real world, however, ensure that she has been declared healthy by your veterinarian, and had her first set of vaccines at least 7 days before any meeting.* See Box 5.3 for a timeline of social development in puppies.

2. What to do: During this period, ensure that your puppy is introduced, in a controlled situation, to as many new people, places, dogs, etc. as you can. Do this carefully – ensure that any dogs that she meets during this time window are well known to you, and are healthy and well-behaved. Ensure that her experiences during this time are positive, happy ones – do not allow your puppy to be bullied, frightened, hurt or overwhelmed during these interactions (or you risk instilling fears in your dog, and accomplishing the opposite of what you hope for). Introduce her gradually to things she will eventually experience, such as visits to the veterinarian, using lots of treats and praise to build positive associations.

3. Use all available resources: Attend a well-run puppy class led by reputable instructors, in which all puppies must provide proof of good health prior to participating. In their book *Puppy Start Right*, Ken and Debbie Martin refer to puppy class as "vaccination against future behavioral problems." In addition, check out resources on socialization such as Sophia Yin's book, *Perfect Puppy in 7 Days*, Ian Dunbar's book *Before and After Getting Your Puppy*, or Ken and Debbie Martin's *Puppy Start Right*. (See the "References and additional resources" section at the end of this chapter for more information on these books.)

*Summarized by Lindell, E.M. (2016) in Clinician's Brief.

Box 5.2. Goals for socializing your puppy

1. Teach your puppy to enjoy the presence of people — first the family, and then friends and strangers, especially children and men.

2. Teach your puppy to enjoy being hugged and handled (restrained and examined) by people, especially by children, veterinarians, and groomers. Specifically, teach your puppy to enjoy being touched and handled in a variety of "hot spots," namely, around his collar, muzzle, ears, paws, tail, and rear end.

3. Teach your puppy to enjoy giving up valued objects when requested, especially her food bowl, bones, balls, chew toys, garbage, and paper tissues to prevent resource guarding and aggression towards humans later in life. (Dunbar 2004).

How can we teach our dogs to enjoy these things? We can do this by introducing new experiences and people gradually, and accompanied by positive experiences like food treats, play, and abundant praise, in order to build up positive associations in our dog's mind.

---

are easier to habituate to than others: it is very difficult to habituate to noises, events, and so on that are associated with negative outcomes. It is for this reason that it is so important to control early introductions between your dog and new types of stimuli. I am unlikely to have an easy time getting my dog to habituate to children, for example, if I allow children to tease her, pull her tail, or otherwise frighten her when she is young or unfamiliar with children. In fact, I run the risk of instilling a fear of children. Ensure that experiences are positive, and that your puppy is not overwhelmed or frightened.

On the other hand, this does not mean that you should "wrap your puppy in cotton wool" (as my mother used to say) and try to protect it from even the slightest chance of any type of stress. Recent research into the concept of "resilience" to stress suggest that early stressful experiences, when present *but very mild and of short duration*, can actually help an animal deal with stressors later in life (Parker and Maestripieri 2011). The developmental outcomes for animals exposed to early life stress follow a U-shaped curve: no exposure to stress, or high levels of early life stress, both result in higher vulnerability to stress later in life (the upper arms of the U), whereas a small amount of early life stress results in the most resilience to later life stresses (the bottom of the U; Parker and Maestripieri 2011). This is part of socialization — introducing your puppy to new people, situations, and objects in a safe, controlled way, so that even if she is initially wary, she will soon learn that these are good things. Given what we know about the importance of being able to exercise some control over the environment in maintaining good welfare, it is likely important to allow the puppy to exercise some choices during socialization, allowing her to approach new people, and so on, at her own pace, and retreat when she wishes.

Figure 5.1: Socialization is crucially important for your dog. (Photo: K.D. Setiabudi)

> **Box 5.3. Enrolling your new puppy in Puppy Class – a great investment for you and your dog!**
>
> Puppies can start puppy socialization classes as early as 7–8 weeks of age. Puppies should be healthy and receive a minimum of one set of vaccines *at least 7 days prior to class*.
>
> 1. One of the best ways to socialize your puppy is through Puppy Class. In a well-organized puppy class puppies will have the opportunity to meet and become familiar with a number of people and dogs in a safe environment.
>
> 2. Puppy Class is also an excellent opportunity for dogs to express appropriate play behavior and learn bite inhibition. Other puppies will teach your dog that if they bite too hard, they will not play along.
>
> 3. To find a good Puppy Class, look for the following:
>
>   a. A positive reinforcement-based trainer, preferably a certified pet dog trainer (CPDT).
>
>   b. Puppies must be required to have up-to-date vaccinations and classes should be held on easy to clean surfaces.
>
>   c. Classes that offer puppies the opportunity to play together off-leash.
>
>   d. Classes that offer short training sessions interspersed with play sessions.
>
>   e. Puppies should be encouraged to explore and manipulate a variety of items like tunnels, chute, toys, and other stimuli.
>
> 4. To find a Puppy Class in your area, check out the APDT Dog Trainer Search at www.apdt.com. See also Box 5.12, later in this chapter, on choosing a trainer.

A final point is that desirable habituation may require maintenance. For example, a puppy who has habituated well to the proximity of cats may still react to them when older, if he spends many years without ever seeing a cat. We'll talk more about how to go about gradually habituating your dog to a stimulus when we talk about systematic desensitization, in Chapter 7 on "Canine behavioral problems and solutions" (in particular, how to proceed when your dog has already displayed a negative reaction to that stimulus).

Socialization is the process of becoming familiar with all kinds of animals, people, places, and things; as well as learning how to behave in society. All puppies need socialization regardless of breed, type, or temperament.

(Ian Dunbar, 2007, The Importance of Early Socialization)

Associative learning and how to make it work for you

There are two types of associative learning that are particularly useful in understanding and in working with our dogs: classical conditioning and operant conditioning. The first of these,

> ## Box 5.4. Socialization timeline: sensitive periods* for socialization
>
> *3 weeks–5 weeks:* This is the primary socialization period. Puppies should be handled by people from birth-on but in this time puppies are more aware of littermates and show signs of distress when separated. Interactions with conspecifics are important in this period. Social play and bite inhibition training begin with the mother and littermates. Pups reared without access to other dogs show marked deficits in their responsiveness to other dogs.
>
> *6 weeks–12 weeks:* This is the secondary socialization period. This is a very important time in social development and proper introductions to all things unfamiliar are crucial to developing a happy dog. This is the time that most puppies bond and learn to live with humans. Puppies leave their mother and become familiar with people, other animals, and their environment. Puppy classes should be attended as soon as the puppy has received its first set of vaccinations.
>
> *12 weeks–6 months or later:* This is the juvenile period in which dogs further develop social and physical skills. This is an important time to train bite inhibition, continue socialization, and begin positive reinforcement training to prevent adolescent issues.
>
> *Sources: Fox 1970; Serpell and Jagoe 1995.*
>
> *Sensitive periods are times in development in which behaviors and preferences are acquired more readily than at other times.

classical conditioning, involves reflex-like, involuntary or emotional responses, and occurs when a stimulus that was previously meaningless ("neutral") to the dog takes on the power to elicit such a reflex-like response. Classical conditioning was introduced to the world in the late 1800s by the Russian scientist Ivan Pavlov, who was studying salivation in dogs in response to food. When presented with the sight and smell of food, hungry dogs will begin to salivate; this is a physiological, reflex reaction to do with breaking down food for digestion. Dogs (or people, for that matter) do not need to be trained to do this. During the course of his research, however, Pavlov's dogs began predictably salivating just *prior* to seeing or smelling the food. Pavlov wanted to understand why and how this phenomenon occurred. He suspected that the dogs had learned to associate the sound of his staff approaching the kennels with the imminent arrival of the food, and so began salivating in anticipation of the food's appearance. By repeatedly and consistently pairing the sound of the bell with the presentation of food, he was eventually able to provoke this same salivation response in the waiting dogs with the bell alone (Figure 5.2). Essentially, he was able to deliberately build the association between the sound of a bell and the food in the dogs' minds. The bell was initially what is called a neutral stimulus; the sound of the bell had no particular meaning for these dogs, until the association had been built in their brains between the sound of the bell and the imminent arrival of food. Once established, the dogs would begin salivating at the sound of the bell, even in the absence of any sight or smell of food. Thus, classical conditioning is the process of building an association between two things in the brain (of a dog, or human, or chicken, etc.); this association can cause reflex-like reactions (anticipation, excitement, salivation) to certain triggers (like the ringing of the bell, for Pavlov's dogs).

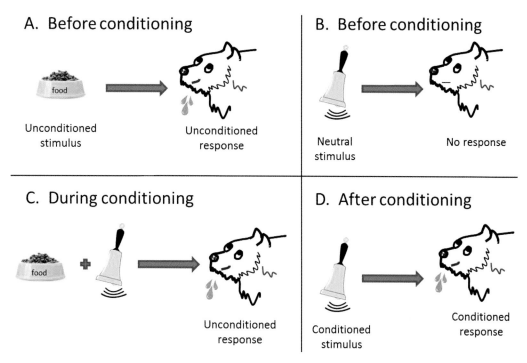

Figure 5.2: Pavlov's dogs and classical conditioning.

Of course, this process happens in humans as well. I remember years ago reading an article in a fitness magazine which advised against eating while watching television. Doing so, the article cautioned, will build up an association between eating and watching television. Eventually, whenever you watch television (whether or not you are actually hungry), you will find yourself craving a snack. I don't watch television much these days, but a similar phenomenon happens to me when I sit down at my laptop to prepare a lecture, read an article, or write; I have always armed myself with a cup of coffee at such times. Now I find that anytime I sit down to work, I immediately want a cup of coffee, even if I just finished one only moments before. I find it hard to get much done without coffee at my side. This is classical conditioning (perhaps augmented by my caffeine addiction); for me, sitting down to work at my desk causes the reflex reaction of physically craving a cup of coffee.

Practical applications of classical conditioning (hint: there are treats involved)

There will certainly be times when we want to be *sure* to build a positive association between our dog and something or someone (i.e., some stimulus or trigger) in the dog's environment. Children, unfamiliar adults, the veterinarian, other dogs, and cats are common examples of this. One way to increase the chances that your dog will view such people and animals as a good thing is to do some classical conditioning. In short (like Pavlov ringing his bell just prior to food), your job is to ensure that whatever the trigger is, it becomes a predictor of good things. So, each time your puppy sees a child approaching on a walk, speak in a happy voice and offer him a delicious treat. Essentially, you are trying to teach him that "children mean good things are going to happen." If he becomes overly excited by the sight of a child or new person (and after you've taught him a "sit" cue), you can ask for a calm "sit" before giving him the treat. This last suggestion gets us into the area of operant conditioning, which we'll talk about next, but the important point is that the sight of the trigger should always predict good things. These common sights, sounds, and smells in his environment should make him feel happy, not fearful.

Operant conditioning: by actively working to control the outcome of a given behavior, we can influence whether or not that behavior continues to be seen

Another very important form of associative learning is known as operant conditioning, which Burch and Bailey (1999) called "the functional relationship between environmental events and behavior." I used to describe this to my animal behavior students as "learning based on consequences" – whether or not an animal repeats a behavior depends largely on what the consequences were for that animal the last time it performed that behavior. Behaviors that are followed by positive outcomes (i.e., are rewarded or reinforced in some way) are more likely to be repeated; behaviors that are followed by negative consequences (i.e., result in some form of punishment) are less likely to be repeated.

Figure 5.3: Dog training is operant conditioning in action. (Photo: K.D. Setiabudi)

Box 5.5. Frequently asked questions about use of rewards in training

What rewards should I use during training?

With rewards, it is important to consider what your dog finds truly rewarding (which may or may not be the same as what you are offering). What we call "high-value" food rewards are a safe bet (see below), but you may need to try different types of treat before finding one for which your dog is really willing to work. For some dogs, the answer may not be food – some dogs may prefer a favorite toy, a quick game of catch, or similar.

Bear in mind, it is a good idea to have a "gradient" of rewards, from a relatively low-level reward (e.g., verbal praise, a piece of dry kibble) for successfully completing an easy task; all the way up to a high-value reward (e.g., pea-sized chunks of hot dog, liver, or cheese) for completing a new or challenging task.

Is it really ok to use food?

Yes! – studies have shown that food is a highly effective way to train behaviors in animals, and food rewards have been used by many successful professional trainers for many years. Be cautious, of course, using food with resource guarders or in groups of dogs.

If I use food, will my dog only respond if he knows I have food?

Not if you are using good training technique! In most cases, the treat should be out of sight until your dog has completed the requested behavior, and once your dog has learned the cue you can begin to "back off" on treats for that behavior. In other words, once he knows the behavior, you can switch from a reward every time to a reward every other time, and so on.

In 1938, the scientist B.F. Skinner published his landmark text, *The Behavior of Organisms*, introducing operant conditioning as "the mechanism through which behavior changes during the lifetime of the individual."[2] An important word in that definition, for those of us who wish to train dogs, is "mechanism" – by actively working to control the outcome of a given behavior, we can influence whether that behavior continues to be seen, increasing in frequency (in the case of a desired behavior), or whether the behavior stops or declines in frequency (in the case of an undesirable behavior). The process of operant conditioning is divided into four quadrants, two of which increase the likelihood of seeing the behavior again, and two of which decrease the likelihood of seeing the behavior again (Figure 5.4). Any behavior that a dog performs can be either *reinforced* (meaning, made more likely to be seen again, and with increased frequency) or *punished* (meaning, made less likely to be seen again, or seen with decreased frequency). In both reinforcement and punishment, the process can be *positive* (+) or *negative* (-). In this case, positive and negative do not mean "good" and "bad" (although they are frequently misinterpreted in this way), but rather positive and negative in the mathematical sense: positive reinforcement and positive punishment both involve *adding* something to the situation; negative reinforcement and negative punishment both involve taking something away. So, in the case of positive

---

[2] The B.F. Skinner Foundation, www.bfskinner.org

Figure 5.4: The four quadrants of operant conditioning (R+, R-, P+, P-). By controlling the consequences of a dog's action, we can control what he learns from the experience. By definition, as well as in practice, rewards (*reinforcement*) following a behavior will increase the likelihood of seeing that behavior again; *punishment* following a behavior will decrease the likelihood of seeing that behavior again.

reinforcement, we are adding something good (i.e., a reward) in response to a desired behavior. In the case of positive punishment, we are adding something bad (i.e., a punishment) in response to an undesired behavior. Similarly, negative reinforcement involves taking away something bad in response to a desired behavior (bearing in mind that, if using this approach, we frequently would be required to apply "something bad" before we could take it away, such as pressure on a choke chain). Negative punishment involves taking away something good (a favorite toy, our attention, etc.) in response to an undesired behavior. In some cases, we (as trainers, owners, etc.) may be administering the reward or punishment (and this is one of the foundations of training), but this isn't always the case. Often, we need to look at the situation and the outcomes of the behavior from the dog's perspective, to determine whether the dog was in fact rewarded or punished in some way for performing the behavior. This is often crucial in understanding why certain behaviors persist, despite our best attempts to get them to stop.

"From a behaviorist's perspective, each day is made up of a series of behaviors that are either reinforced or not reinforced." Burch and Bailey (1999), How Dogs Learn.

Operant conditioning has been used in animal training for many years now,[3] beginning in earnest in the 1940s and 1950s with Skinner's group and the trainers Bob Bailey, Marian and Keller Breland.

---

[3] Burch and Bailey (1999) provide a concise history of dog training throughout the ages.

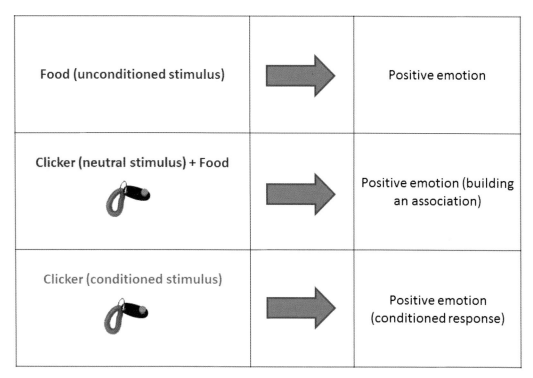

Figure 5.5: How clicker training works (see also Figure 5.2, on classical conditioning).

Operant conditioning is the basis of "clicker training," the rewards-based training method originally developed by Bob Bailey (although B.F. Skinner used something resembling a clicker in his earlier work in operant conditioning). Clicker training was popularized and expanded on by Karen Pryor in the 1980s with her wonderful book, *Don't Shoot the Dog*; Dr. Pryor's organization now provides extensive resources and training for those interested in positive-reinforcement-based and clicker training methods[4] (Figure 5.5). The clicker acts as a "conditioned reinforcer" – so, *once it has been associated in the dog's brain with good things* (e.g., a highly desirable food reward), the sound of the clicker can be used to reinforce (reward, increase the frequency of) a given behavior. The clicker is also referred to as a "bridge" or "secondary reinforcer"; the dog must first learn what the sound of the clicker means ("good job! you are about to get a nice reward!"); once she understands this tool, the clicker becomes a "bridge" between action and reward, and an important way to communicate *immediately* with your dog. One important aspect of using rewards in dog training is the speed at which the dog is rewarded for successfully performing the behavior. If the reward takes longer than a second or two to reach him, he may not understand exactly why he has received the reward (or, he may have performed a different, less desirable behavior in the meantime). The behavior immediately preceding the click is the behavior you are rewarding – so, good timing is essential. As a "bridge" between action and consequence, the

---

[4] For general resources on clicker training, check out http://www.clickertraining.com/; for more in-depth information on learning to train using rewards-based methods (including certification programs), visit https://www.karenpryoracademy.com/

clicker can enable instant marking and reinforcing of the desired behavior. Lack of a click is also communication: no click means no reward (because the correct behavior was not performed). This ability to provide immediate feedback and reinforcement to your dog can greatly improve the effectiveness of your training.

Similarly, the British veterinarian and behaviorist Ian Dunbar began publishing on and popularizing "lure-reward training," and training of puppies, in the 1980s. In 1993, Dr. Dunbar founded the Association of Pet Dog Trainers (APDT) in the United States; APDT is now one of the most widely-recognized training organizations in the western world. Like Dr. Pryor, Dr. Dunbar also provides a wealth of resources on humane, rewards-based training.[5] More recently, British researcher and author John Bradshaw has become an articulate spokesperson for humane treatment of companion animals. Dr. Bradshaw's book, *Dog Sense*, provides a wealth of science-based information on domestic dogs, and is a wonderful resource for dog owners everywhere.

Dogs go with what works ... for them!

From a practical standpoint when it comes to training, the primary question for most of us who live and work with dogs is: what *really* works? An additional question, for those of us who love dogs (including you, if you're reading this book), is: which training method will result in the happiest, most well-adjusted dog? To accomplish the most in our busy schedules, with the greatest commitment to the well-being of our canine companions, we recommend always using methods based on the considerable scientific research into how dogs learn, and how dogs react to different training and management styles. In a sense, always make "evidence-based training decisions" – which type of training has been examined, tested, and proven both effective and the most humane option, time and again?

There is considerable scientific evidence for effectiveness and speed of rewards-based (including food) training. Whether or not to use food in training has often been a point of disagreement in dog training circles, but food is highly motivating for the vast majority of dogs, and the appropriate use of food rewards has been repeatedly shown to be a fast and effective way to train (e.g., Feuerbacher and Wynne, 2014; Fukuzawa and Hayashi, 2013; Okamoto et al. 2009). Food is an incredibly powerful training tool with most dogs – use it! For more on which rewards to use, see Box 5.5, and Box 7.6 in Chapter 7. Rewards-based training is often referred to as "positive training" or "using positive methods," but these terms are confusing and potentially misleading (given "positive" punishment). Another term for this type of training is "force free" (highlighting what trainers using these methods do *not* rely on, i.e., physical force)[6]. Rewards-based training, however, does not mean lax, lenient, or overly permissive training. Dogs need boundaries, just like children (and adults, for that matter). Rewards-based trainers rely more on training and

---

5 Many of Dr. Dunbar's resources are available through his website, http://www.dogstardaily.com/
6 Do not, however, rely solely on terms used by trainers when selecting a trainer for your dog, as anyone can claim to use these types of training. Instead, evaluate the trainer's methods for yourself, to ensure no harsh physical punishments will be used with your dog.

consistently reinforcing good behavior, and avoid the risks associated with positive punishment (more on these risks later in this chapter). It's important, however, to be clear and consistent about where the boundaries are, and not forget to reward good behavior! As Diana Bird notes in her essay on Victoria Stillwell's "Positively" blog:

> It's fine to give a dog a lot of freedom, but first we have to prepare them for that. Initially we have to control their environment a lot and teach them what is expected … Set them up to be right; teach the rules/expectations/skills and as they succeed, give them more independence. If we just give a dog (or child, or anyone) total freedom from the beginning, they will do what works for them and that may not work for us.[7]

As we've said before, we do believe it is important to respect dogs as dogs, and not place unrealistic expectations on their abilities and behavior (particularly if we haven't taught them how we wish them to behave in the first place). However, that being said, for dogs to live successfully and safely in human households, certain basic rules apply, and some types of behavior are inappropriate and even dangerous. It is our responsibility to train our dogs to be upstanding members of canine society (or, hire someone to train them for us, at least to the point where they can coexist with humans and other animals happily and safely).

Applications: "How do I get him to stop doing that?!?"

Going back to our discussion of the four quadrants of operant conditioning, all of which represent ways in which animals learn: which of the four should we be using if we wish to embrace progressive, humane training? Positive reinforcement is an obvious choice; reward behaviors that we want to see again. But, how should we go about stopping behaviors that we don't want to see again, if positive punishment is not recommended? The most commonly recommended approach is to use negative punishment; remove the rewards for the behavior (in other words, make it *not* work for the dog). For example, if your dog jumps up on you during greetings, think about why he does this – most likely, he's excited, wants to interact with you, and wants your attention fully focused on him. You may or may not have reinforced this behavior at some point by giving him this attention, petting him, and so on (we are often guilty of rewarding this behavior when the dog is a puppy, only deciding that the behavior is no longer desirable when our puppy is no longer quite so small and manageable). To remove the reward (i.e., your attention) for this behavior, you can pointedly look away and turn your body away from the dog, or even leave the room if the behavior persists. Only interact again with your dog when he is sitting calmly (or at least, not jumping up on you). To increase the speed at which he gets this message, teach him a "sit" command first; he cannot simultaneously sit and jump up on you. Then, when he greets you or anyone else, you can ask for – and reward! – the sit behavior. Teaching a preferred, alternate behavior is giving the dog a way to succeed in a given situation (e.g., by teaching him to sit when greeting humans), and is an important way to reinforce desirable behavior in your dog. I

---

7   Bird, D. 2015. *The problem with the "Pack Leadership" mentality*. Victoria Stillwell, Positively.com. Accessed October 27, 2015.

encourage my students to teach their dogs "sit to say please" at a young age; teach the dog that, anytime she wants something from you, the way to get it is to sit calmly (vs. jumping up, whining, barking, etc.). "Sit to say please" is a great skill for any dog, as it provides you a way to encourage and reward calm behavior in almost any situation, and provides her a way to communicate with you that there is something she wants.

Positive reinforcement (to encourage a behavior), and negative punishment (to stop a behavior) combined with teaching an alternate (preferred) behavior, are the most humane forms of operant conditioning to use in training

There are a number of fantastic training resources to help you train both basic and advanced behaviors, and to reduce unwanted behaviors, in your canine companions. We've included some of our favorite training resources in Box 5.6. In the meantime, here are some other

> ### Box 5.6. Just a few of our favorite training books, DVDs, and websites!
>
> **Books**
>
> Donaldson, Jean. *Train Your Dog Like a Pro.* (includes training DVD). Howell Book House (Wiley).
>
> Martin, Ken and Debbie Martin. *Puppy Start Right.* Karen Pryor Clickertraining/Sunshine Books.
>
> McConnell, Patricia. *Family Friendly Dog Training: A six-week program for you and your dog.* McConnell Publishing, Ltd.
>
> Pryor, K. *Don't Shoot the Dog! The new art of teaching and training.* Ringpress Books.
>
> Tillman, Peggy. *Clicking With Your Dog: Step-by-step in pictures.* Karen Pryor Clickertraining/Sunshine Books.
>
> Yin, Sophia. *Perfect Puppy in 7 Days: How to start your puppy off right.* CattleDog Publishing.
>
> **DVDs**
>
> Donaldson, Jean. *Perfect Paws in 5 Days.* Perfect Paws Productions. (Includes useful "troubleshooting" tips for common training mistakes.)
>
> McConnell, Patricia. *Lassie Come!* McConnell Publishing, Ltd. (Dr. McConnell offers a whole series of booklets and videos on training and dealing with behavioral problems on her website; this is just one example.)
>
> **Websites**
>
> Karen Pryor Clickertraining: www.clickertraining.com
>
> Kikopup's YouTube Channel (Emily Larlham's library of free training videos): https://www.youtube.com/user/kikopup
>
> Eileen and Dogs (*training, body language, and more*): http://eileenanddogs.com/
>
> Learning About Dogs, with Kay Laurence: www.learningaboutdogs.com

important terms and concepts to be familiar with, when adding tools to your toolbox for working with your dog:

Extinction

A behavioral response can be reduced or eliminated, if the conditioned stimulus or cue is repeatedly presented without any consequences (i.e., without any accompanying reward or punishment). So, for example, one way to reduce a dog's learned behavior of begging at the table is to consistently ignore this behavior – that is, ensure that the behavior is *never* rewarded. A related phenomenon is the "extinction burst" – initially, the ignored dog may actually increase the behavior to try and reiterate what she wants (by barking, whining, etc.). For extinction to happen, you must hold strong, continuing to ignore the behavior until it stops.

Shaping

One way to train a new behavior, especially a fairly complex one, is to break down the desired behavior into tiny increments or steps, and reinforcing the dog *at each incremental step* until you've got the full behavior from your dog. This is how many dogs used in films and television are trained to do complex behavioral sequences, but can work for basic behaviors like focusing on you, going to his bed, and so on. So, for example, you might start the process of shaping "come when called" by clicking and rewarding your dog for just looking towards you and/or taking a step in your direction. Eventually, you require him to move closer and closer to you before rewarding him, and finally only rewarding him for returning all the way to your side.

Box 5.7. Teaching bite inhibition

Bite inhibition is one of the most important things your dog can learn in order to safely interact with people. Bite inhibition training is how we teach our dogs to safely use their mouths (what level of pressure is OK). It is not always possible to avoid a dog bite but dogs that have developed bite inhibition will be far less dangerous if ever pushed to the point of biting.

One way that we teach bite inhibition is by allowing our dog to mouth us in play. When his teeth create any sort of pressure, respond with a high pitched, "ouch!" and end the play session. Your dog may be shocked or confused by your behavior but once he is calm, you can resume play. If he exerts pressure with his mouth again, repeat the sequence. Over time your dog will learn that we humans have pretty thin skin and that they need to be careful.

Puppies also learn bite inhibition in playing with littermates. When play becomes too rough, the play session stops. In this way they learn how to play fair. Dog expert Patricia McConnell* recommends against rough-housing with your dog, as fun as that may seem to some, as this may overexcite the dog or encourage rough play with humans (that sadly, sometimes results in injury).

*McConnell, P. (2008) Play Together, Stay Together. McConnell Publishing Ltd.

Capturing

Essentially, this process of training a new behavior involves waiting for the dog to voluntarily perform the behavior, and then instantly rewarding it. Timing is crucial – if you wait too long between the behavior and your reward, the dog may not know why she is being rewarded (this is where a clicker becomes a very useful tool!). Of course, you may "set up" the situation to encourage the dog to perform the behavior, but the dog should be deciding for herself to do a behavior. So, for example, one way to train "sit" is to wait for the dog to sit on her own, and then immediately reward her for the sit. Repeat this process, eventually adding a cue (a verbal "sit" cue, or perhaps a hand gesture), so that she will sit reliably when asked.

Reinforcement schedules

We noted above that, to be most effective, rewards should be given immediately (within a second or so) following the desired behavior. Rewards can be given on more than one "reinforcement schedule," however. Continuous reinforcement means that *every* time the dog performs the behavior, he earns a reward. Continuous reinforcement is the best schedule to use when teaching a dog a new behavior. Intermittent reinforcement, on the other hand, means that the dog does not receive a reward every time he does the behavior, but only periodically, and either on a fixed schedule (e.g., every other time he does the behavior) or on a variable schedule (unpredictably; e.g., on the third, fourth, eighth, and tenth time). Don't switch from continuous to intermittent reinforcement too soon, however. Intermittent reinforcement should only be used once the dog has successfully learned the behavior, but intermittent reinforcement is often the best way to maintain the acquired response (i.e., the dog will continue to work in anticipation of the reward, like a slot machine player in a Las Vegas casino, even when the reward doesn't happen each and every time).

Learning a new behavior is fastest when the reward is immediate, of high value to the dog, and continuous

Management

This is not technically part of training, but rather an important first step. Management in dog terms means managing the dog's environment to increase safety (for the dog, or for other people and animals potentially interacting with the dog), reduce triggers or distractions for your dog, and limit the opportunities for the dog to repeat behaviors that you don't want to continue. It does not solve the problem (like a good training or behavioral modification program can), but it does lower risk, increase effectiveness of training, and reduce damage due to misbehavior. So, for example, if you have a dog who growls at unfamiliar visitors to the home, the first response should be a management one: put the dog in a comfortable crate, or quiet room, when visitors are expected, and keep the dog safely on leash any time he is around strangers outside of the home. Your next steps should be to address the problem through training and behavioral modification; we'll talk more about this type of situation in Chapter 7.

Figure 5.6: Our dogs rely on us not to put them in harm's way. Avoid use of force and harsh punishments when working with your dog, as these techniques damage the relationship between you and your dog, and run the risk of making the behavior worse, not better. (Photo: iStock)

A note on crates and crate-training: I am a big believer in the value of teaching any dog to be calm and comfortable in a crate or kennel. Many canids will seek out dens for shelter and safety; it is a natural instinct in many dogs. If crate-trained correctly, many dogs view their crates as their "safe place" (or, as Sophia Yin termed it in her crate-training guidelines,[8] their "personal bedroom"). They will voluntarily retreat there when they are tired or stressed, and can be calmly contained in a crate for an extended period of time. For tips on how to crate-train your dog in the most humane and effective way, see Box 5.8. This is not to say that there is not a risk of overusing the crate! Crates are an excellent tool to use during house-training and other aspects of puppy training, as a way of managing the environment (i.e., containing the dog) during behavioral modification work, and are often necessary containment for dogs when traveling, being boarded, or at the veterinarians. However, dogs were not designed to be contained in a crate for 8 to 10 hours a day (or more), every day, while owners are at work or out of the house. So, our advice is: do crate-train your dog. In my experience, the vast majority of dogs can be easily taught to tolerate their crates, to the point where they will seek them out on their own as preferred sleeping spots. But, don't abuse this tool; in most cases, extended containment in a crate should be a temporary situation (although you can certainly allow your crate-friendly dog free access to his crate, by simply leaving the door open, for him to seek out on his own schedule).

The risks of positive punishment

Today, rewards-based training is considered the state of the science in dog training. Older methods that rely heavily on force and positive punishment are on the wane. The dangers of positive punishment (increased fear, high risk of increased aggression, damage to human–animal bond) are well documented in the scientific literature on dog training methods and veterinary medicine. Positive punishment is one of the four quadrants of operant conditioning, and when

---

8   Yin, S. 2009. *Low Stress Handling, Restraint, and Behavior Modification of Dogs & Cats.* p. 126.

> ## Box 5.8. Teaching your dog to love her crate
>
> At its heart, the key to crate-training is classical conditioning – teach your dog that great things (such as delicious treats) happen when she is in her crate, and bad things don't happen in the crate. Don't ever punish your dog when she is in her crate. If you want or need to contain your dog in her crate as a "time-out" for any reason, that's fine, but be careful how you get her into the crate – direct her calmly, using the usual cues and rewards, into the crate. Don't drag or physically force your dog into the crate. She shouldn't ever view being put into her crate as a punishment. There is also an aspect of systematic desensitization to crate-training; the process of getting them used to the crate should be a gradual one. Following this protocol, most dogs will come to accept their crates within a week. Dogs fearful of containment may take longer (and require more patience on your part).
>
> Begin by feeding your dog her usual meals at the door to her crate; secure the door open to ensure that she does not get hit or frightened by a swinging door. Once she is visibly comfortable with this, move her bowl *just* inside the door of the crate. Once she is comfortable with this arrangement, gradually move the bowl, in increments, to the very back of her crate. Leave the door open throughout the process, and periodically (between meals) toss a few treats (or a stuffed Kong, or similar) into the crate for her to discover on her own.
>
> Once she is completely comfortable eating in her crate, you can begin closing the door briefly while she is eating; gradually increase the time that the door remains closed. Ask her to sit before letting her out of the crate. If she does whine or bark when closed in the crate, you can try simply asking for the sit and not letting her out until she is quiet. If this does not solve the problem, try going back a few steps and repeating the steps above. Train (using food rewards) a cue that means, "go into your crate." Keep a comfortable bed or mat in the crate, and leave the door to the crate open throughout the day, so that she can use the crate whenever she wants to. Take common sense precautions when leaving any item in the crate with your dog, however – don't leave anything that presents a choking or obstruction hazard. And, if you discover that your dog panics in the crate, despite following the steps above, you may need to think of an alternate way to contain her.
>
> *Adapted from: Yin, S. (2009) Low Stress Handling, Restraint and Behavior Modification of Dogs & Cats.*

used correctly, carefully and humanely, can be effective under some circumstances and with some individuals; however, use of positive punishment runs substantial risk of having unintended side effects, and lends itself to abuse or misuse, particularly in the hands of an inexperienced or uninformed trainer. Physical damage due to such techniques is reported periodically in the news, and described in the veterinary literature. For example, in 2013, Grohmann and her colleagues described a tragic case of a young (1-year-old) German Shepherd dog who was essentially strangled by his own owner. The owner admitted to disciplining the dog by hanging the dog by his choke chain off the ground for approximately 60 seconds. This is obviously an awful, worst-case scenario; however, incidences like this are sadly not all that uncommon when looking into the reports of consequences of force-based punishment techniques like choke chains. The techniques are common enough to have names in certain dog training circles: 'hanging' or (in the case of swinging the suspended dog by his leash) 'helicoptering.' Harsh, force-based punishment techniques such as these are not training; they are abuse, and have no place in our relationship with our dogs.

"Fear doesn't train a reliable dog" Ian Dunbar (2006), in Rafkin (2006).

Perhaps more pervasive, but less obvious to the untrained eye, are the emotional, behavioral consequences of using harsh, force-based techniques when training a dog. Use of electric shock collars, for example, are often associated with negative behavioral and emotional outcomes in dogs; in an often-cited 2004 study on police dogs trained using shock collars versus those trained without use of shock, researchers Schilder and Van der Borg noted that the dogs trained with shock collars showed more stress-related behaviors (such as avoidance, lowered body posture, yelps and squeals) in the presence of their trainer than dogs who had never received electric shocks. As the authors note, the dogs trained with shock collars "evidently have learned that the presence of their owner (or his commands) announces reception of shocks, even outside of the normal training context. This suggests that the welfare of these shocked dogs is at stake, at least in the presence of their owner."[9] Force-based, "confrontational" training techniques often result in an increase in fearful and aggressive behaviors from the dog, including aggression directed towards the owner; this is documented repeatedly in the scientific literature on domestic dogs (Herron et al. 2009; Blackwell et al. 2008; Rooney and Cowan 2011). As Ian Dunbar[10] notes, "Fear doesn't train a reliable dog." In addition, rather than improving the situation, use of positive-punishment-based training techniques is associated with a decline in the bond between owner and dog, and can contribute to the dog's eventual relinquishment by the owners (Deldalle and Gaunet 2014; Kwan and Bain 2013).

In using positive punishment to try and quickly "stop" a behavior, you may wind up suppressing the symptom of a problem, without actually treating the problem. There is a training saying that comes across my Facebook page periodically: "Don't punish the growl." Dogs use growling as a way to communicate that they are uncomfortable in a situation; if this communication is ignored and the perceived "threat" continues, the dog may escalate to the next level of response (often involving some form of aggression). And, we as humans don't get to decide what the dog perceives as a threat – if it scares him, it scares him, regardless of whether we think he should be afraid of the happy toddler, friendly vet tech, and so on. So, by simply punishing a dog for growling without addressing the reasons he is growling, we may create a "ticking time bomb": the next time the dog is in that situation, he may not growl at all, but may go right to the bite. This is a situation that generally ends badly for everyone involved, including the dog. Far better to immediately remove the growling dog from the situation, identify the reason(s) for his fear and/or aggression, and use scientifically-supported techniques to change the dog's emotional reaction to this trigger, and thus his future behavior in the presence of the trigger. We'll talk more about dealing with serious behavior problems such as aggression in Chapter 7.

In addition, positive punishment is surprisingly difficult to administer effectively. For punishment to work, it must be administered during or *immediately* following the undesirable behavior; it must be administered *consistently* (i.e., the behavior should be punished *each and every time*

---

9 Schilder, M. and Van der Borg, J. 2004. Training dogs with help of the shock collar: Short and long term behavioural effects. *Applied Animal Behavior Science* 85: 319–334.
10 In Rafkin, L. 2006. *The Anti-Cesar Millan: Ian Dunbar's been succeeding for 25 years with lure-reward dog training; how come he's been usurped by the flashy, aggressive TV host?* SFGate.com article collections, 10-15-2006; accessed 9/15/2010.

it occurs, and the dog should not be left alone in a position where it can perform the behavior without receiving punishment); and the punishment needs to be sufficiently serious to leave a lasting impression on the recipient. We as humans are rather infamous for administering punishment for canine misbehavior well after the fact. When I was very young, the accepted method of house-training puppies involved leaving them unattended during the day while owners were at work or out of the house, and upon returning home to the (almost inevitable) puppy mistake on the carpet, rubbing the puppy's nose in or very near the site of his transgression, while punishing him verbally or physically, and then taking him outside to "show him where he should be going." This is an example of positive punishment administered after the fact (possibly hours after the fact), well beyond the time interval during which we can be sure that the puppy understands the association between behavior and outcome. This association is necessary for learning to occur. For this reason (along with a number of other reasons), this method of house-training a puppy is now recognized as highly inefficient. Current house-training methods involve controlling the puppy's environment (with a suitably comfortable crate, for example), watching him carefully to learn the signs that he needs to go outside (to keep him from "practicing" the behavior of eliminating inside the house), allowing him ample opportunities to meet his toileting needs outside in an appropriate location (setting him up for success), and rewarding him each and every time he successfully eliminates outside (reinforcing the behavior we want to see again). I have used both old and new house-training approaches in my life with dogs (the first when I was young and my family knew no better). I have found the second, more modern approach much easier, quicker, and more effective (and as a result, much less frustrating). For step-by-step house-training tips and troubleshooting (for both young and adult dogs), I highly recommend Patricia McConnell's wonderful booklet, *Way to Go! How to housetrain a dog of any age.*[11]

Remember the Five Freedoms as goals for a happy dog

Going back to our initial discussion of the Five Freedoms (in Chapter 2) as a way to assess our dogs' quality of life, if we are using techniques that create or perpetuate fear in our dogs, we are not meeting the dogs' basic need as described in the fifth Freedom: the Freedom from Fear and Distress. If we regularly use techniques designed to cause pain, such as hitting, pinch collars, or electric shock collars, we are additionally not meeting the third Freedom: the Freedom from Pain, Injury and Disease.

Importantly, training techniques using only rewards-based, force-free methods have been found to be as effective as the force-based techniques, *without the associated behavioral and welfare risks,* and can foster increased attentiveness towards the owner and enhanced learning of subsequent cues and behaviors (Cooper et al. 2014; Deldalle and Gaunet 2014, Rooney and Cowan 2011). Rewards-based, force-free methods have been used for decades with overwhelming success by experts and trainers like Pryor, Dunbar, Bradshaw, and others. Always remember the goal: a successful living relationship between the dog and his or her human family. There are

---

11 Available through Dr. McConnell's website (www.patriciamcconnell.com) and major online booksellers.

multiple methods out there promising the achievement of this goal; it is up to us, as the ones with the greatest control over the situation, to choose methods which safeguard the physical and emotional welfare of our canine companions. Given the established risks associated with force-based methods, and the proven benefits of using rewards-based methods, the choice seems clear; when training, use rewards-based methods that are effective, humane, and that allow you to provide your dog with the best quality of life, as outlined by the Five Freedoms. I remember vividly the first dog I trained using clicker training, instead of the older methods I had used with previous dogs. One thing that struck me was how different this training experience was than the previous ones: with clicker training, my dog was enthusiastic about learning and eager to please, and clearly enjoyed our training sessions, as evidenced by his behavior and body language. In addition, he learned new cues and behaviors with impressive speed (and as much as I loved him, I wouldn't categorize him as unusually intelligent by canine standards). And no, training with the clicker did not require me to carry a clicker and treat bag everywhere I went with my dog, from that day forward. Once he learned the behavior with the help of the clicker, he continued to repeat it successfully (with only periodic rewards provided). Other colleagues have reported noticing similar differences in the dog's response to training when using rewards-based versus force-based "traditional" training methods. Choose methods that will ensure a happy dog. For some readily-available training tools and helpful products popular with many progressive trainers, see Box 5.9.

So, is it ever ok to say "no" to your dog while training? Many trainers use something called a "no reward marker" (meaning, a signal to the dog that the behavior they have just performed will not be rewarded), citing improved communication with their dogs, but others firmly state that it is better to simply ignore the mistake by the dog, go back to the starting point and reissue the command.[12] One recent study trained 27 dogs to complete a simple task; half the dogs only received positive reinforcement for correct responses (mistakes were ignored); the other half received both positive reinforcement for correct response, and a tone used to indicate when they made a mistake. The researcher found that the dogs trained using only positive reinforcement did considerably better. These dogs learned more quickly, and reached a significantly higher level of proficiency, than the dogs who received the "no reward marker" telling them when they were wrong in addition to the rewards for correct responses (Rotenberg 2015). Even if you decide to use a "no reward marker" in your training, it is important to remember that there is a big difference between calmly (or cheerfully, in what is sometimes called a "happy interrupter") saying, "Uh oh!" or "Oops" when your dog fails to succeed; and yelling at him, hitting him, or tapping on an electric shock collar. And if, like me, you may occasionally yelp (loudly) in surprise when your dog, calmly sleeping next to you on the couch, suddenly bursts into a volley of barking, accept that you are human. Humans make mistakes and we may sometimes yell at our dogs. You can gauge how aversive your yelling is to your own dog by his behavior – does he glance at you in mild surprise and continue his behavior, or freeze and cower in fear? Some trainers suggest using

---

12 For more on use (or not) of no reward markers (NRMs), check out Denise Fenzi's post on this topic, at https://denisefenzi.com/2014/05/01/behavior-chains-part-10/, or Susan Garrett's comments on NRMs, at http://susangarrettdogagility.com/tag/nrm/ (accessed 4/20/2016).

Box 5.9. Some helpful training tools

By now, you may have gathered that certified behaviorists do *not* recommend harsh training tools, such as electric shock collars, pinch or prong collars, or choke collars. There are, however, some very helpful training tools available that are endorsed by the majority of behaviorists and force-free trainers. Here are a few of our favorites:

Head halters

These give better control over the dog's head, meaning better ability to control where he is looking/what he is focused on. These are very useful in working with reactive dogs, as well as dogs who tend to pull when on-leash. Correct fit is important, so be sure to purchase the correct size and follow fitting instructions.

- The Gentle Leader (PetSafe, Knoxville, TN)
- The Halti (The Company of Animals, Chertsey, Surrey, UK)

Body harnesses

Although these may not be quite as effective as a head harness at controlling the dog's focus, front-clip harnesses can be useful for dogs who will not tolerate a head harness for any reason, and can be helpful for dogs who pull (standard back-clip harnesses – unlike the Freedom harness mentioned below – are not generally recommended for use with pullers or reactive dogs). As with head halters, correct sizing and fitting is important.

- The Freedom No-Pull Harness (modified back clip design with optional front clip; 2 Hounds Design, Monroe, NC)
- The Positively No-Pull Harness (Victoria Stillwell Positively.com)
- The Easy Walk Harness (PetSafe, Knoxville, TN)

Other useful tools (various manufacturers)

- Long-line or training leads: these leashes, normally about 20 feet in length, are very useful for training recall and other behaviors that will eventually be done off-leash. Note that these are *not* the same as extendable/retractable leashes, which are not recommended by the vast majority of behaviorists and trainers (primarily for safety reasons).
- Martingale collars: these collars are designed for dogs whose necks are bigger than their heads, and who can easily slip out of a standard flat buckle collar (greyhounds, for example). They tighten only slightly when pulled, providing a less risky, more humane alternative to the choke chain.

training to make this event, if it happens regularly, less aversive for your dog. For example, train your dog that your startled yelp means, "come sit at my side and I will reward you for doing so".

All of this begs the question: if rewards-based methods are so superior, why is there still so much resistance in some sectors of the dog training world to relinquishing the older, force-based techniques? In the US, popular culture plays a role, with celebrity television hosts and online "experts" espousing outdated methods relying on positive punishment (while simultaneously, on-screen warnings often caution viewers "not to try these techniques at home," begging the question of how safe, or even useful, some of the techniques are). Some of this persistence of "traditional" training methods may simply be the natural human resistance to change; change can indeed be scary, particularly if we believe our income relies on the continued use of these methods. Perhaps more insidious is the appearance of a "quick fix" when using positive punishment to stop a behavior; punishing a behavior in the moment may, and often does, result in the behavior stopping *in that moment*. However, whether the behavior will be repeated in the future, or in the absence of the punisher, or whether any of the other adverse side effects of positive punishment (increased fearfulness, defensive aggression, etc.) will be seen, are crucial questions. The research consistently demonstrates that the occurrence of unintended and/or unanticipated results of using positive punishment is a very real possibility.

Perhaps the most troubling suggestion for why force-based training techniques persist is the idea that inflicting punishment may in fact be rewarding in some way for the punisher. When we are angry or frustrated or embarrassed, lashing out at those responsible can sometimes act like a pressure vent, relieving some of these unpleasant emotions. In discussing methods for dealing with misbehavior in young human children,[13] Dr. Jed Baker stresses that, "realizing (that) our children's challenging behaviors are not threats to our own competence, but instead are a function of the youngsters' tenuous ability to cope with frustration" is an important part of changing these behaviors for the better (Baker 2008, p. 32). Rather than getting angry and venting our frustration and embarrassment on our dogs, far better for us to recognize that, when our dogs do not behave as we would like, this misbehavior may be due to our dog's physical, cognitive, or emotional inability to respond appropriately. This inability may be due to being overexcited or "over threshold" and therefore unable to control her own actions and emotions, or it may be due to her lack of understanding what a given cue means or what is expected of her in a given situation. Recognizing this can help soothe our anger, and focus us back on the important work at hand: identifying the problem and associated triggers, and coming up with a plan to change the behavior for the better. Dogs are not humans, but as our understanding of the cognitive and emotional abilities of dogs continues to increase, changes in dog training techniques parallel, increasingly, changes in treating humans with behavioral problems. As in dog training, many techniques regularly used to address mental health and behavioral problems in humans in the past (such as isolation, ice water baths, overuse of physical restraints, and drastic surgical procedures) seem horrific to us today. Similarly, many of the harsh, punishment-based techniques used to train dogs in the past are now considered unnecessary, inhumane, and inappropriate by modern training professionals.

The "D" word: why won't the "dominance myth" die?

The concept of "dominance" as applied to domestic dogs has in recent years been the most inflammatory word in dog training circles, to the point where recommendations from leading experts in dog behavior on choosing a trainer often advise dog owners to avoid any trainer who even uses the word when discussing reasons behind canine misbehavior.[14] In the science of animal

---

13 Baker 2008. *No More Meltdowns: Positive strategies for managing and preventing out-of-control behavior.*
14 See, for example, the American College of Veterinary Behaviorists' online "Guide for Choosing a Trainer" http://www.dacvb.org/wp-content/uploads/How-to-select-a-trainer-A-guide-for-owners.pdf

behavior, what does the term dominance actually mean? Historically, dominance has been defined as a relationship between individual animals, or an animal's role within a social group, which is usually established by force and/or aggression by some and submission by others. Most importantly, dominance in this sense determines who has priority access to resources such as food, mates, and territory. Strict dominance hierarchies were believed to exist in wild wolf packs, and as wolves are cousins and (in a sense) precursors of today's domestic dogs, many argued that domestic dog behavior could be interpreted by studying the behavior of wolves. In the more recent past, some dog trainers have claimed that much of domestic dog misbehavior is in fact a struggle for dominance between dog and human. In order to solve these problems, therefore, these trainers claimed it was necessary to reassert dominance over our canine companions, using the method that dogs understand best: physical force. Techniques such as forcibly rolling the dog on his back, physically restraining him on his side, growling in his face, and so on, were recommended as ways to reassert dominance over the dog.

There have been a number of elegant rebuttals in the scientific and professional literature to these popular misconceptions of what the concept of dominance means for domestic dogs, as well as clarifications about the social dynamics of wild wolf packs (they are not as rigidly hierarchical as the original studies, based on groups of unrelated wolves living in captivity, believed) and the huge influence of the process of domestication on domestic dog behavior (domestic dogs have evolved over countless generations to cooperatively, not competitively, coexist with humans). Even in free-ranging feral and "village" dogs (unowned and unsupervised dogs living in close association with human settlements), data on whether or not dog groups form stable hierarchies among themselves is mixed, perhaps due to the high levels of individual variation seen in dogs and their impressive ability to adapt to local conditions. When hierarchies are seen, they tend to be maintained by submissive behaviors offered by one member of the interaction, rather than overt aggression or threats displayed by the other. Authors like the behaviorist and scientist John Bradshaw are careful to note that dominance is a quality of relationships, but not necessarily a motivation for the dogs in *creating* those relationships, and likewise not a quality of an individual dog (Bradshaw et al. 2009). The American Veterinary Society of Animal Behavior's (AVSAB) position statement on "The use of dominance theory in behavior modification of animals" clarifies the terms, provides answers to common questions, and cautions about the dangers of misuse of force in animal training.[15] Box 5.10 summarizes some of the key points in arguments against the use of the concept of dominance in dog training.

As a final thought on the persistence of the "dominance myth" in dog training, in her essay (appropriately titled "Why nobody will ever agree about dominance in dogs") the British researcher Carri Westgarth points out two factors that likely contribute to this issue. First, in scientific circles, the idea that dominance is a fixed attribute of an individual animal or position within a social group (i.e., animal social groups are organized into fixed, rigid hierarchies, and some animals are "dominant" while others are "submissive") may very well reflect outdated ideas about non-human animals as not capable of any form of reason, and about how animals make the decisions necessary to live their daily lives. We are continually learning just how much non-human animals

---

[15] Available online at www.AVSABonline.org

Box 5.10. Three reasons why "dominance" is not a great approach to living with dogs

Domestic dogs are not wild wolves

Interpreting dog behavior based on wolf behavior is risky at best. And besides, some of the fundamental ideas about wolves that form the basis of the misapplication of dominance theory today have been misinterpreted, or even been proven false by wolf researchers. Take the "alpha roll," the practice of forcibly rolling the dog on his back and holding him there. This was originally claimed to mimic a dominant or "alpha" wolf physically asserting dominance over a subordinate wolf. In fact, what really happens among wolves is that the subordinate wolf rolls on his back voluntarily as a way of communicating submission to the alpha wolf (our dogs sometimes do something similar). There is a big difference between voluntarily communicating compliance with a group member, and being forcibly restrained in an uncomfortable and vulnerable position; the latter is much scarier, and more likely to evoke defensive aggression.

Most "misbehavior" by domestic dogs is not about dominance at all

Dominance, in animal behavior terms, is about who gets priority access to resources, and is generally not written in stone. In other words, an individual who gets best access to resources in one circumstance may not get best access to a different resource, and/or in a different set of circumstances. In the majority of domestic issues between our dogs and ourselves, dominance is not the issue; rather, the dog is performing a behavior that has worked for him in the past (either because we have allowed the behavior to persist, haven't trained him an alternate behavior, or have inadvertently rewarded the behavior). For example, a dog who steals food from the kitchen counter is not trying to assert his dominance over his human companions; the most likely scenario is that the dog has successfully obtained a highly desirable food item in this location in the past, and so is highly motivated to repeat the behavior. Rather than using this to communicate dominance over his human companions, he may not even make the attempt in the presence of humans, particularly if he has been scolded or punished in the past for this behavior. By allowing the dog access to unattended food items on the counter, we allow the behavior to persist and continue to "work" for the dog.

Aggression breeds aggression

Trainers who still subscribe to the "dominance myth" often rely heavily on physical force to "reassert dominance" over the dog. Studies have repeatedly shown that owners using positive punishment and "confrontational" training techniques are more likely to experience worse behavior from their dog, even owner-directed aggression, than owners using only rewards-based training (e.g., Herron et al. 2009). If you behave in a way that your dog perceives as a direct threat to his well-being, you are more likely to provoke defensive aggression in your dog. This is a primary reason why organizations such as the American College of Veterinary Behavior and the American Veterinary Society of Animal Behavior recommend avoiding trainers who advocate an approach based on physically reasserting "dominance" over your dog.

are capable of, and we now know that there is a lot more cognitive processing and emotional activity going on in most animals' brains than we have previously given them credit for. In other words, things are a lot more complicated than we thought, and whether or not a dog succeeds in a social group (and, humans are part of a domestic dog's social group) does not depend on simple

Figure 5.7: Our dogs rely on us to safeguard their welfare; one way to do this is to work with the most up-to-date, humane trainers and training methods. (Photos: K.D. Setiabudi)

dominance or lack thereof. Interactions between dogs, and between dogs and their humans, are complex and changeable. The idea that any one individual is "dominant," that the hierarchy is fixed, is likely a gross oversimplification.

A second point that Dr. Westgarth makes involves the popular use of the term "dominant," and the fact that many dog owners are using the term in a subtly different way from the scientific community – not dominance as in "all-powerful 'leader of the pack' in a rigid linear hierarchy," but dominance as in "getting the behavior that I want and expect from my dog." Some dogs, like some humans, may be stronger or "pushier" than others, but this does not necessarily mean that every move they make is motivated by the desire to dominate others, or that, to change this behavior, we need to firmly reassert our authority over these dogs. As Dr. Westgarth puts it:

> Simply by the fact that my own dogs (mostly) do as they are asked instead of defying or challenging my authority, I am in layman's terms dominant over them. However, again this word is used to describe the *outcome* of our interactions, not the *input* of how we got there; in my case by teaching them that it is worthwhile doing as I ask.[16]

As Burch and Bailey observed in 1999, "a great deal of the popular literature related to training dogs is full of mythology and pseudoscience." This observation still holds true to a frustrating degree today. The wide reach of the Internet, in particular, allows anyone to distribute information about dogs and dog training. It can be very difficult for an Internet user to distinguish between information based solely on the opinions and/or misunderstandings of the individual or group posting the information, and information supported by solid science, conducted by a wide range of researchers studying learning, canine cognition, and behavior, and "field-tested" by hundreds of professional, humane trainers. Everyone has (and is entitled to) their opinions, but when it comes to working with our dogs, surely we should be basing our actions on the most

---

16 Westgarth, C. 2016. Why nobody will ever agree about dominance in dogs. *Journal of Veterinary Behavior*. (*emphasis added*)

> **Box 5.11. An all-too-common mistake we make in training our dogs**
>
> The other day I was walking my own dog in my neighborhood and saw a scene which I've seen played out far too many times before: in this case, two dogs (one bouncy white German shepherd dog in the lead) escaped from their owner's grasp and dashed across a road to investigate another dog. Their owner immediately chased after them, calling loudly and, when they did not respond to his calls, angrily for them to return. Eventually, when the dog that had caught their attention was removed by its owner, the two dogs turned and made their way back to their owner. With the help of a friend, their owner managed to grasp the collar of the shepherd, yelling at the dog and smacking him sharply on the head as punishment for ignoring his initial calls.
>
> Setting aside emotional reactions to the owner's behavior here, and while sympathetic with his frustration and/or concern over his dogs crossing a road, what is wrong with his response, from a learning perspective?
>
> Essentially, by punishing this dog immediately after the dog's return to the owner (which was after all, what the owner wanted), the owner is teaching the dog, "when I do get hold of you, I will punish you." This is unlikely to motivate the dog next time to return promptly. Far better to use learning theory and rewards-based training to train a really strong, reliable "come when called" or recall behavior – train the dog that, when they respond promptly to your call, they will get a really great reward! Consider motivations for your dog's behavior – can you provide something more rewarding than whatever is tempting him away from your side? Initially, this may mean setting aside your irritation at slow or inconsistent responses, but keep the end goal in mind; done correctly, training a solid recall is essential for happy (and safe) coexistence with your dog. Many trainers suggest training *two* "recall" cues; a standard cue for everyday use, and an "emergency" cue (for when you *really* need compliance, and which is *always* rewarded with a high value reward).

accurate information available, information supported by the considerable body of research in this area, combined with our own gut instincts as to what is humane and in the best interests of our canine companions. As the behaviorist and trainer Kathy Sdao notes in her wonderful essay "Organic Training,"[17] trainers who rely on positive reinforcement-based methods focus first on building a relationship of trust with their dogs, and commit to the principle, "First, do no harm." Throughout this book, we have attempted to provide readers interested in going further into a given topic or technique with resources and references from reliable sources and recognized experts in the fields of canine behavior and rewards-based, force-free dog training.

The principles of learning described earlier in this chapter are essential to successfully and humanely training dogs, but we should not forget that, like people, every dog is an individual. As important as it is to understand and apply the principles of learning and rewards-based training, it is also important to understand the individual dog with which you are working. You know your dog, probably better than any other human on the planet, and you can use your understanding of what makes her tick as part of your training program. What motivates your dog? What treats

---

17 Available at www.kathysdao.com/articles/organic-training/ (accessed May 2016).

or activities does she value (and hence will work for) above all else? Are there things that reliably distract or frighten your dog, making it difficult for her to learn appropriate behavior in certain situations? And, can you avoid these triggers while training, or (better yet) design ways to help your dog habituate to these triggers? We will talk about how to use learning as part of a plan to treat behavior problems in Chapter 7 on "Canine behavioral problems and solutions," but in the following section we will present some practical tips to help you get the behaviors that you want from your canine companion, while reducing or eliminating unwanted behaviors.

As we noted in the introduction, this book is not meant to be an exhaustive text on any one aspect of dog husbandry, and this includes training. There are many excellent books, DVDs, websites and training courses which do a comprehensive, effective job of helping you learn how to train your dog, or to improve your existing training skills (and we've listed a few of these in Box 5.6). Box 5.9 lists some useful training tools that you may wish to add to your training toolbox. What follows are a few "training basics" that may be helpful to those readers who have not done much training in the past, or who are looking to change or improve the way that they train.

Training tips to get you started

Set realistic expectations

As Linda P. Case noted in a recent blog post,[18] there is no perfect dog. Herding dogs love to (and will) herd things, most dogs shed hair everywhere they go, some dogs may not be comfortable with children or strangers, and yes, many dogs do indeed love to eat poop. Acknowledge that your dog is an individual, and has limitations (but is no less deserving of love because of this). Training may go slowly; you may need to manage your dog's world for her safety and the safety of others; improvements may be small, at least initially; and relapses may occur. Take it one day at a time, and instead of focusing solely on the "finished product," make the goal to help your dog's life be a bit calmer and easier each day – small incremental changes will add up! This isn't to say that we shouldn't try to improve our dog's behavior, particularly if that behavior is in any way a safety issue. But, we may need to get help from qualified professionals, be patient, be persistent, and keep expectations realistic.

Be proactive, not reactive

I had the good fortune that the first dog training class I ever attended (as a student and dog owner, and many years ago now) was with a wonderful progressive, professional trainer. The first thing she said to the group was a game changer for me, and so has always stayed with me. She looked us all in the eyes and said firmly, "Stop yelling 'No!' at your dog." Instead, she instructed us that, when faced with a behavior we didn't like, we should first think about *what we would rather the dog was doing in that moment*, or in that situation. Then, *train that behavior*, and reward

---

[18] "The Perfect Dog," on *The Science Dog* blog, https://thesciencedog.wordpress.com/. March 10, 2016.

the heck out of them when they do it successfully! Today, I think of this as my "golden rule" – rather than endlessly reacting to misbehavior, and railing against our dog's shortcomings, better to take a proactive approach; decide what you want the dog to do, and train the dog to do just that. Similarly, be sure you are using learning processes appropriately, rewarding the behavior you want, and not inadvertently punishing a dog for complying with your wishes. See Box 5.11 for a very common mistake dog owners make – don't be that person![19]

Remember the rules of reinforcement

Remember our earlier discussion about the best (and fastest) way to train a new behavior, and to maintain an existing behavior:

- To train a new behavior, reward the behavior immediately, consistently, and with a sufficiently high-value reward (with higher-value rewards used when the behavior is novel or challenging).
- Once your dog knows the behavior (so, can and will correctly respond to the cue at least nine out of 10 times), you can start to switch from a continuous reinforcement schedule to an intermittent one. Don't completely stop rewarding your dog for doing as you ask! Although ideally you will always mark a successful response with a quick verbal "yes!" or "good job," gradually scale back on how frequently he gets the higher-value rewards (but be sure to continue to reward the behavior periodically, or you risk the extinction of the behavior).

Be consistent!

Most of us have very busy lives these days, and it can be hard to maintain "consistency of message" with our dogs. For example, if you don't want your dogs to get up on the furniture or beg at the table, then the furniture should *always* be off limits, and the dog should *never* receive treats directly from the table. If, in a weak moment or on a particularly chilly winter's evening, we allow our dogs to stretch out next to us on the couch, we risk confusing him the next time we scold him for climbing up on his own to this very comfortable sleeping spot. Decide in advance what the rule is (and how you will reinforce an alternate behavior, such as sleeping on his dog bed rather than the couch), and be sure that everyone in the dog's household is working from the same rule book. Similarly, good communication is important: be sure to use clear, consistent cues when working with your dog; don't change the cues (or the rules) day to day or you'll confuse your dog and make your own job harder. Dogs as a species do not use verbal language, and although dogs can be trained to understand the meaning of some words, many dogs respond better to gestures than verbal prompts; you may want to add gestures as cues when training your own dog. Keeping a few verbal cues (such as "sit," "freeze," and "come") in your repertoire is a good idea, though, for times when your dog can hear but not see you.

---

19 Also check out "Control unleashed" by Leslie McDevitt, available at http://controlunleashed.net/

Set your dog up for success

Training generally works best if you begin training without distractions present; once your dog understands the cue and is able to reliably respond correctly, you can begin to introduce distractions. This doesn't mean suddenly moving your training sessions from the quiet of your living room to the dog park at peak hours; small increases or baby steps are better (as many trainer friends say, "you have to go slow to go fast").

Short training sessions (5 to 15 minutes at a time), done regularly (at least once a day, if you can manage it) are much more effective than trying to do a 2-hour training marathon on a weekend day. Dogs generally lose focus after about 20 minutes, and lessons learned repeatedly become more of a habit.

Similarly, don't ask for too much, too fast! If you want to train your dog who hates nail trims to tolerate nail trims, for example, you may need to start by just gently and briefly touching one of her paws (and rewarding her when she allows you to do this); over a series of sessions, *gradually* increase your interactions: so, for example, gradually work up to being able to hold one paw for 15 seconds or so, then to being able to hold her paw while simply holding the clippers (so that she can see them, but they are not touching her), then to just touching her paw briefly (not trimming) with the clippers, then to trimming just *one* nail before ending the session, and so on … and always, always reward her for tolerating each increase in interaction; don't punish her if she doesn't succeed at a given level, just give her a break, then start again at the previous level, until she is comfortable (use your skills reading body language to assess when your dog is comfortable at any given level of interaction, before "raising the stakes"). I have seen this process work, numerous times; for some dogs (especially those who have never had a painful experience with nail trims) this process can go quite quickly; for others (who have had a bad nail trim experience at some point, and so have learned that nail trims really can be painful) this can take months. Be patient, and work at your dog's pace.[20]

Make a point sometimes to spontaneously reinforce a preferred behavior

When you come across your dog doing something you really like (even if just sitting calmly in his bed while you are working, talking on the phone, or whatever), make a point to verbally praise him, and offer him a small tasty food treat. Note that this goes against some more rigorous training philosophies (which recommend never rewarding a behavior unless you have asked for it), but I think for family dogs, rewarding them periodically for good behavior even when you haven't asked for it helps to maintain these behaviors (and, a happy dog).

---

20 In some cases, you also could try changing the situation! For example, try switching from traditional nail trimmers to use of a Dremel pet nail trimmer (as with any new, potentially scary item or situation, introduce the Dremel gradually and in conjunction with lots of food treats, and praise for calm behavior; follow the instructions on using the Dremel on dog nails.) For more on this process of systematic desensitization, see Chapter 7.

Use management wisely

As we talked about earlier, use management strategies to manage your dog's environment, to reduce risk, enhance learning, and reduce opportunities to practice inappropriate behavior (after all, practice makes perfect, and we really don't want our dogs to perfect any risky, damaging, or annoying behaviors). Don't become complacent, however; remember that management alone won't solve a serious behavior problem. Behavioral modification and training will be required (more on treating behavior problems in Chapter 7).

Don't hesitate to get professional training help for you and your dog!

Figure 5.8: To ensure a happy dog, use humane, effective, force-free training methods based on the science of how dogs learn. (Photo: ©Donna Kelliher, www.DonnaKelliher.com)

Training is a technical skill that must be learned; no matter how much you love your dog, don't expect to know intuitively how to train him. There are many wonderful trainers out there who have taken the time to learn the necessary skills and to get certified by respected organizations like APDT or the Karen Pryor Academy (KPA), and who have years of experience to draw upon in working with any dog. However, not all trainers are created equal (so to speak), and it is important to do your homework before taking your dog to a trainer. As noted earlier, a number of well-respected organizations have put together handouts and information sheets, available online, with recommendations for choosing a trainer for your own dog.[21] Box 5.12 lists some of the most common suggestions for choosing a trainer.

Your dog relies on you not to put him in harm's way

Most important, however, is to follow your own best instincts, as a caretaker and advocate for your dog – ask a potential trainer about his or her training philosophy, what tools (collars,

---

21 Three examples are the American Veterinary Society of Animal Behavior's position statement on choosing a trainer, available at http://avsabonline.org/resources/position-statements; the Association of Professional Dog Trainers page on choosing a trainer (with additional links to information behind their recommendations) at https://apdt.com/pet-owners/choosing-a-trainer/; and the American College of Veterinary Behaviorists' handout, available under "Resources" on their website at http://www.dacvb.org/

> **Box 5.12. Commonly recommended strategies for choosing a trainer (adapted from AVSAB's "How to Choose a Trainer").**
>
> *1. Do your research beforehand:* Realize that dog training is a largely unregulated industry; anyone can call themselves a trainer, or even a behaviorist, regardless of whether they have any real training themselves. Look for evidence of certification from a reputable program, such as the letters CPDT-KA (Certification Council for Professional Dog Trainers, Knowledge-Assessed); or KPA-CTP (Karen Pryor Academy Certified Training Partner), among others. These letters are a great guideline, but not a guarantee – you should still follow steps 2 and 3, below.
>
> *2. Look for trainers who use rewards-based training methods, supported by the science of learning and canine behavior:* Avoid trainers who recommend use of force-based, positive-punishment-based methods like "alpha rolls," scruffing or choking, tools like pinch collars or choke chains, or insist on use of electric shock collars for training.
>
> *3. Look for trainers who are helpful and respectful – to both you and your dog!* Trainers should clearly explain what they are doing, allow sufficient time for practicing new behaviors, and be willing to work individually with class members to help them succeed. If they behave towards you or your dog in any way that makes you uncomfortable, look for another trainer.

harnesses), and so on, he or she recommends, and ask to observe a class before committing. If you are uncomfortable with anything the trainer is asking you to do to your dog, find another trainer.

Take-home messages

1. Understanding the scientific principles of how dogs learn will be hugely helpful to you in understanding your own dog's behavior, in training new behaviors in your dog (and stopping unwanted behaviors), and (if necessary) in being an informed consumer when selecting professional training help for your dog. Feel confident in the path that you have taken: there is a substantial body of excellent research supporting the effective and humane use of learning principles in dog training.
2. Use rewards-based training to teach your dog that doing what you ask is worthwhile, as it will earn her good things in life. Be proactive in training your dog to do what you want by clearly communicating expectations and rewarding success, and by removing the reinforcement for behaviors that you don't want. Don't jeopardize your relationship with your dog by using harsh, punishment-based methods or tools, and don't allow others to do so with your dog.
3. Use management when necessary; manage the dog's environment to improve safety, reduce opportunities for misbehavior, and help control the outcome of a given behavior. Remember: dogs go with what works!
4. Don't hesitate to enroll your puppy in a good "Puppy Class" and/or seek professional training help when necessary, but select this help with care, following the recommendations of professional veterinary and training organizations, and your own instincts.

References and additional resources

Baker, J. (2008). *No More Meltdowns: Positive strategies for managing and preventing out-of-control behavior*. Arlington, TX: Future Horizons, Inc. 150 p.

Blackwell, E.J., Twells, C., Seawright, A., & Casey, R. (2008). The relationship between training methods and the occurrence of behavior problems, as reported by owners, in a population of domestic dogs. *Journal of Veterinary Behavior: Clinical Applications and Research* 3: 207–217.

Bradshaw, J. (2011). *Dog Sense: How the new science of dog behavior can make you a better friend to your pet*. New York: Basic Books. 324 p.

Bradshaw, J.W.S., Blackwell, E.J., & Casey, R.A. (2009). Dominance in domestic dogs – useful construct or bad habit? *Journal of Veterinary Behavior* 4: 135–144.

Burch, M.R., & Bailey, J.S. (1999). *How Dogs Learn*. Hoboken, NJ: Howell Book House/Wiley Publishing Inc.

Case, L.P. (2014). *Beware the Straw Man: The Science Dog explores dog training fact and fiction*. Mahomet, IL: Autumn Gold Publishing. 189 p.

Cooper, J. J., Cracknell, N., Hardiman, J., Wright, H., & Mills, D. (2014). The welfare consequences and efficacy of training pet dogs with remote electronic training collars in comparison to reward based training. PLOS One, 9(9), e102722. doi:10.1371/journal.pone.0102722

Deldalle, S., & Gaunet, F. (2014). Effects of 2 training methods on stress-related behaviors of the dog (*Canis familiaris*) and on the dog–owner relationship. *Journal of Veterinary Behavior: Clinical Applications and Research* 9: 58–65. doi:10.1016/j.jveb.2013.11.004

Donaldson, J. (2010). *Train Your Dog Like a Pro*. Hoboken, NJ: Howell Book House/Wiley Publishing Inc.

Dunbar, I. (2004). *Before and After Getting Your Puppy: The positive approach to raising a happy, healthy and well-behaved dog*. Novato, CA: New World Library. 224 p.

Dunbar, I. (2007). The Importance of Early Socialization. *Dog Star Daily*, Wed, 03/28/2007. http://www.dogstardaily.com/training/importance-early-socialization. Accessed March 1, 2016.

Feuerbacher, E.N., & Wynne, C.D. (2014). Most domestic dogs (*Canis lupus familiaris*) prefer food to petting: Population, context, and schedule effects in concurrent choice. *Journal of the Experimental Analysis of Behavior* 101(3): 385–405.

Fox, M.W. (1970). Overview and critique of stages and periods of canine development. *Developmental Psychobiology* 4: 37–54.

Fukuzawa, M., & Hayashi, N (2013). Comparison of 3 different reinforcements of learning in dogs (*Canis familiaris*). *Journal of Veterinary Behavior* 8(4): 221–224.

Grohmann, K., Dickomeit, M. J., Schmidt, M.J., & Kramer, M. (2013). Severe brain damage after punitive training technique with a choke chain collar in a German shepherd dog. *Journal of Veterinary Behavior: Clinical Applications and Research* 8(3): 180–184. doi:http://dx.doi.org/10.1016/j.jveb.2013.01.002

Hart, B., Hart, L.A., & Bain, M.J. (2006). *Canine and Feline Behavior Therapy* (2nd ed.) Ames, IA: Blackwell Publishing. 373 p.

Herron, M., Shofer, F., & Reisner, I. (2009). Survey of the use and outcome of confrontational and non-confrontational training methods in client-owned dogs showing undesired behaviors. *Applied Animal Behaviour Science* 117: 47–54.

Kwan, J., & Bain, M. (2013). Owner attachment and problem behaviors related to relinquishment and training techniques of dogs. *Journal of Applied Animal Welfare Science* 16: 168–183.

Lindell, E.M. (2016). Top 5 puppy behavior tips. *Clinician's Brief*, February 2016.

Martin, K.M., & Martin, D. (2011). *Puppy Start Right: Foundation training for the companion dog*. Waltham, MA: Karen Pryor Clickertraining/Sunshine Books.

McConnell, P. (2003). *Way to Go: How to Housetrain a Dog of any Age*. Black Earth, WI: McConnell Publishing, Ltd. 23 p.

McConnell, P. (2005). *Lassie Come! How to get your dog to come every time you call*. (DVD). Black Earth, WI: McConnell Publishing, Ltd./Tawser Dog Videos. 45 min.

McConnell, P. (2006). *Family Friendly Dog Training*. Black Earth, WI: McConnell Publishing, Ltd. 108 p.

McConnell, P. (2008). *Play Together, Stay Together*. Black Earth, WI: McConnell Publishing, Ltd. 90 p.

McDevitt, L. (2007). *Control Unleashed: Creating a focused and confident dog*. South Hadley, MA: Clean Run Productions, LLC. 226 p.

Okamoto, Y., Ohtani, N., Uchiyama, H., & Ohta, M. (2009). The feeding behavior of dogs correlates with their responses to commands. *Journal of Veterinary Medicine and Science* 71(12): 1617–1621.

Parker, K.J., & Maestripieri, D. (2011). Identifying key features of early stressful experiences that produce stress vulnerability and resilience in primates. *Neuroscience and Biobehavioral Reviews* 35: 1466–1483.

Pryor, K. (2002). *Don't Shoot The Dog! The new art of teaching and training*. Surrey, UK: Ringpress Books. 202 p.

Rafkin, L. (2006). *The Anti-Cesar Millan: Ian Dunbar's been succeeding for 25 years with lure-reward dog training; how come he's been usurped by the flashy, aggressive TV host?* SFGate.com article collections, 10-15-2006; accessed 9/15/2010.

Rooney, N., & Cowan, S. (2011). Training methods and owner-dog interactions: Links with dog behavior and learning ability. *Applied Animal Behaviour Science* 132: 169–177.

Rotenberg, N. (2015). Training a new trick using no-reward markers: Effects on dogs' performance and stress behaviors. Master's Thesis, Hunter College, New York. Available from CUNY Academic Works. http://academicworks.cuny.edu/hc_sas_etds/12

Schilder, M. & Van der Borg, J. (2004). Training dogs with help of the shock collar: Short and long term behavioural effects. *Applied Animal Behavior Science* 85: 319–334.

Serpell, J., & Jagoe, J.A. (1995). Early experience and the development of behaviour. In J. Serpell (Ed.), *The Domestic Dog*. Cambridge, UK: Cambridge University Press. pp. 82–102.

Tillman, P. (2000). *Clicking with Your Dog: Step-by-step in pictures*. Waltham, MA: Sunshine Books. 209 p.

Westgarth, C. (2016). Point-counterpoint: Why nobody will ever agree about dominance in dogs. *Journal of Veterinary Behavior* 11: 99–101.

Yin, S. (2009). *Low Stress Handling, Restraint and Behavior Modification of Dogs & Cats*. Davis, CA: Cattledog Publishing. 469 p.

Yin, S. (2011). *Perfect Puppy in 7 Days: How to start your puppy off right*. Davis, CA: Cattledog Publishing. 175 p.

Chapter 6

Canine mental health: the importance of enrichment

Photo: iStock

WHAT CONSTITUTES GOOD MENTAL HEALTH IN DOGS? WE TEND TO THINK OF the ideal companion dog as being calm, stable and unflappable, happy and friendly. As in humans, these characteristics would seem to reflect good mental wellness. Happiness, the focus of this book, is certainly conducive to maintaining optimal mental wellness. But, as we touched on in other sections of this book, what does it mean to be happy, from a dog's perspective? And how can we be sure that our own dogs are happy? The fact is, we don't really know what happiness feels like for a dog, but we can certainly infer (from their behavior, body language, facial expressions) whether they are experiencing a positive, pleasurable emotion (comparable to our happiness), versus a negative emotion (like fear). Furthermore, in living and working with dogs, we can determine (being as objective as possible) whether they are primarily experiencing positive emotions, and only occasional negative ones; or the reverse, experiencing

Figure 6.1: Mental well-being is an important part of good quality of life, for dogs as it is for humans. (Photo: E. Grigg)

primarily negative emotions, with only occasional moments of relief. This is, in one sense, a measure of quality of life.

We would like for our dogs to have the best possible quality of life (see also Chapter 2 for more on assessing and defining quality of life in dogs). Mental wellness is an important aspect of overall good health and quality of life, just as much as physical wellness is. In fact, as we know from many studies done in human medicine, the two are often intricately related. Chronic stress can have negative impacts on physical health, and chronic physical ailments (particularly those causing chronic pain) can take a serious toll on mental wellness. So, in this chapter we will cover what the science tells us about what is important for good mental health and behavioral wellness in domestic dogs.[1] This will be followed, in the next chapter, by what can happen when things go wrong: common behavioral problems in dogs, and the best, science-based recommendations for modifying behavior.

---

1 Issues surrounding physical health will be covered primarily in Chapter 8, on physical wellness, and in Chapter 9, on veterinary care.

Figure 6.2: Two (now retired) residents of the university teaching dog colony. (Photo: P. Turmenne)

What happens when dogs aren't happy? Canine responses to chronic stress

When I began my first job teaching applied animal behavior to veterinary students, the university where I was teaching had a colony of resident teaching dogs: 25–30 mixed-breed dogs (the exact number varied) that were used in teaching non-invasive techniques like the canine physical exam to students. These dogs lived in the on-campus kennels facility for up to two years, after which time they were adopted out to suitable homes (usually students who had become attached to them during their time on campus). As soon as the campus became aware of my background in behavior, however, an alarming number of students came to me to express their concerns about the welfare of the teaching dog colony. Their concerns were not, for the most part, for the physical health of these dogs, who were fed high-quality food and provided with regular veterinary care by the on-campus veterinary clinic. Rather, the students were primarily concerned about the dogs' mental health. At that time, there was no behavioral wellness or enrichment program in place for those dogs.[2] They were taken on short walks, twice a day, but spent the rest of their time in their individual indoor kennels, unless they were needed in a lab on a given day. There was no initial training done with the dogs, to familiarize them with lab procedures and settings, prior to their introduction into the labs. Concern for aggression between dogs (and the risks

---

2 There is now an active behavioral wellness and enrichment program in place at this facility, largely run by veterinary students committed to providing these dogs with the best possible care, and with a full-time kennels facility manager to ensure the ongoing well-being of all dogs. In addition, dogs now spend a maximum of 18 months in the program prior to adoption/retirement into a suitable home.

that such aggression might present to students working with the dogs) meant that the dogs were never allowed to socialize and play with other dogs. Perhaps not surprisingly, the types of behavioral problems often seen in kenneling facilities, such as excessive barking, barrier aggression (barking or lunging at other dogs through the kennel bars), and repetitive behaviors like pacing and spinning, were seen in many of these dogs.

Faced with this situation, I began to look at the literature on maintaining behavioral wellness of dogs housed in kennels, studying work conducted in rescue shelters and research facilities where dogs are housed. I wanted to know what the experts, those professionals and researchers who worked with kenneled dogs on a regular basis, had found about the best ways to maintain good behavioral health and to minimize the behavioral problems often seen in kennels facilities. I quickly discovered that kennels are often very stressful situations for dogs. Stressors present in kennels include high levels of noise caused by multiple barking dogs; confinement to a small area, often with minimal social contact; the lack of exercise and control over their environment; the unfamiliar setting and disruption of normal routines; and in some cases, the loss of their familiar human companions (Hetts et al. 1992; Hubrecht et al. 1992; Beerda et al. 1999, 2000; Stephen and Ledger 2005; Rooney et al. 2007; Hennessy 2013). Studies looking at behavior as well as levels of the hormone cortisol (a physiological indicator of stress) supported the stressful nature of kennels for dogs.

Many of the behaviors students were reporting in the facility matched behaviors reported in the literature and associated with chronic stress in dogs. Although individual dogs do vary in their responses to stress, it is generally agreed that maladaptive and repetitive behaviors, such as self-mutilation and stereotypies, are indicative of chronic stress (Beerda et al. 2000). In addition to physiological changes (e.g., in cortisol levels), other changes commonly associated with the chronic stress of kennel life include indications of frustration (such as chewing, barking, or howling), conflict (body-shaking, paw-lifting), and a lowered/fearful body posture (Hewson et al. 2007).

Many of the studies cited above were looking at ways to provide for the mental health of these dogs, even in the face of the unavoidable stressors of living in a kennels facility. And, importantly, many of these same methods of improving mental well-being can also be beneficial for dogs living in a home. This is particularly true for dogs living in single-dog homes; or with owners who must of necessity be out of the home for long hours at work, leaving the dog home alone; or for dogs housed even temporarily in crates or boarding kennels.

What do dogs need to be happy? Factors influencing behavioral wellness

In 2010, the Association of Shelter Veterinarians in the US published their landmark, *Guidelines for Standards of Care in Animal Shelters*. An entire chapter of this very useful publication[3] is dedicated to improving and maintaining the behavioral health and mental well-being of companion animals living in shelter situations. That document also cites the Five Freedoms (described

---

3 Available for free download at http://www.sheltervet.org/

in more detail in Chapter 2 of this book) as goals for all animals living in shelter situations. Being aware of the common causes of stress for dogs, and what approaches work to minimize and mitigate that stress (even for kenneled dogs living in relatively stressful situations), can be very useful in formulating ways to ensure the very best behavioral health and mental well-being of our own, in-home canine companions.

> *The structural and social environment, as well as opportunities for cognitive and physical activity, are important for all species of animals.*
>
> (ILAR 1996, in ASV 2010)

The ASV Guidelines, summarizing a wealth of research into canine responses to stress, note that the lack of control over their environment is a major stressor for dogs living in shelters. Confinement is stressful for many animals, and the effects of this stress are exacerbated when mechanisms which help the animal to cope with stress (such as places to hide, social companionship, mental stimulation, and physical exercise) are limited or not available. Continued exposure to these stressors can result in chronic anxiety and cause or exacerbate the abnormal behaviors described earlier in this chapter. So, in light of this, what do the ASV Guidelines recommend to ensure optimal mental health and behavioral welfare of these dogs?

> *The purpose of enrichment is to reduce stress and improve well-being by providing physical and mental stimulation, encouraging species-typical behaviors (e.g., chewing for dogs and rodents, scratching for cats), and allowing animals more control over their environment.*
>
> (ASV 2010)

1. *Environment*: Housing should allow the dog to meet its most basic behavioral needs (ASV 2010). Harking back to our earlier descriptions (Chapter 2) of the Five Freedoms, dogs should be provided with:
   a. adequate water and nutritious food (freedom from hunger and thirst, the First Freedom);
   b. adequate shelter from the elements and a comfortable resting area (freedom from discomfort, the Second Freedom);
   c. regular medical care from qualified veterinarians (freedom from pain, injury, or disease, the Third Freedom);
   d. sufficient space (for feeding, resting, moving around, as well as separate areas for urination and defecation) and the company of other dogs, and/or humans (freedom to express normal behaviors, the Fourth Freedom); and
   e. conditions and treatment designed to minimize mental suffering to the greatest extent possible (freedom from fear and distress, the Fifth Freedom).
2. *Enrichment*: Environmental enrichment, which first came to prominence in the zoo world, is defined by the Association of Zoos and Aquarium's Behavior Scientific Advisory Group[4] as "a dynamic process for enhancing animal environments within the context of the animal's

---

[4] The AZA has a great web page with more information on enrichment for a wide variety of species, at https://www.aza.org/enrichment/

behavioral biology and natural history. Environmental changes are made with the goal of increasing the animal's behavioral choices and drawing out their species-appropriate behaviors, thus enhancing animal welfare". In short, we can greatly reduce stress in many animals, dogs included, by providing them opportunities for expressing natural behaviors, and for exercising at least some measure of control over their environments. At the bare minimum, the ASV Guidelines state, animals must be provided with regular social contact, mental stimulation, and physical activity (ASV 2010). Providing enrichment, the authors stress, is as important as feeding and veterinary care, and should not be considered optional. We will talk more about suggestions for enriching the lives of companion dogs later in this chapter.

3. *Daily Routine*: The ASV Guidelines also recommend keeping daily routines on as regular a schedule as possible, as unpredictability can be stressful for many animals. They also suggest including in the daily schedule as many positive experiences (such as feeding time, playtime) as possible, at regular times, as many animals respond with positive emotions to the *anticipation* of these predictable, positive events. I know my own dog is very aware of when her breakfast and dinner times are, and does indeed display positive anticipation of these events, along with other daily favorites such as walk time. Although I suspect that dogs living in less stressful home environments may be more resilient to occasional (and sometimes unavoidable) changes in their owner's schedule, it is interesting to note that a common recommendation for reducing stress in companion dogs with fear and anxiety issues is to institute and maintain a very predictable daily routine.

Enrichment for the companion dog: ways to make your dog's day a lot more interesting

Throughout the life of your dog, providing enrichment is a great way to improve your dog's quality of life. It has also been found to be effective in helping to maintain cognitive function and slow the impacts of aging on a senior dog's brain (Milgram et al. 2004). I used to refer to this with my students as the "Sudoku effect", after the challenging mental number puzzles so popular a few years ago, or as "use it or lose it" syndrome – by engaging their brains (and bodies), we can help our dogs maintain optimal health and quality of life for the longest time possible. Providing enrichment for your dog is a very attainable goal, and can be done for very little cost (even for free, in many cases), and we'll describe some tried-and-true methods in this section. We will also list some popular toys and devices developed specifically to provide enrichment for dogs, ranging in cost from very reasonable to fairly pricey. You can judge for yourself which would appeal to your dog, and which fall within your budget.

Figure 6.3: Dog friends ready for a play session on the beach. (Photo: R. Hack/S. Anthony)

Enrichment: the basics

Social and playtime

One of the most basic forms of enrichment, but one that is both vitally important and easily accessible, is playtime (Box 6.1). For many dogs, this will consist of playtime with other dogs; regular social interaction with other dogs is very important for the well-being of these dogs (and, helps them meet the Fourth Freedom, freedom to express normal behavior). For other dogs, who may be less social or less reliable around other (particularly unfamiliar) dogs, this may consist of playtime with one or two familiar dog friends. All dogs, whether they are highly dog-friendly, or more solitary, wary of or aggressive towards other dogs, will benefit from positive experiences such as playtime with a familiar human. So, if your dog loves trips to the busy dog park, then providing him regular access to this activity is a great, free way to provide enrichment. If on the other hand, this isn't available where you live, or if you have a dog who does not enjoy this (and not all do), you can provide her with the valuable gift of your time – time spent with your dog, playing in the back garden, on walks in the park or around the neighborhood, is another great way to provide her with enrichment that she needs for optimal mental health.

Physical exercise

Another very important form of enrichment for companion dogs is regular physical exercise. Some of this can be provided during the social/playtime described above, and this is a nice advantage for those owners who have a sociable, well-mannered dog who loves to spend a daily session running and playing with other dogs at the dog park. For others, though, we should be dedicating

Figure 6.4: Play is an important part of your dog's quality of life, at all ages. (Photo: © Donna Kelliher, www.DonnaKelliher.com)

Box 6.1. The importance of play

Play is important in the development of communication in dogs. Play, usually in the form of mouthing, begins as early as 21 days and play-fighting begins by 4 to 5 weeks. Dogs develop sensory and motor skills through play and learn to signal the difference between play and aggressiveness. Allowing dogs to play is an important part of meeting the fourth Freedom: Freedom to Express Normal Behavior.

1. The play bow is thought to have a metacommunicative function, communicating the intent to play (Bekoff 1972, 1977, 1995). Play signals can help animals detect that the encounter will be playful rather than aggressive.

2. Early play experience can alter rates of aggressive behavior later in life. In social play, animals reciprocate frequently between winning and losing postures, enabling them to gain experience in these situations while in a safe context (Špinka et al. 2001).

3. Dogs with more play experience may be better able to cope with aggressive encounters. Through play, animals learn to cope with unexpected experiences and this prepares them for aggressive encounters (Špinka et al. 2001).

4. Play behavior is an indicator of well-being and opportunities for play enhance the life of a happy dog (Donaldson et al. 2002).

at least some time each and every day to ensuring that our dogs are getting the physical exercise that they need (and, simply letting our dogs out into the back garden for 20 minutes or so does not adequately meet this need – unless, of course, your back garden consists of a multi-acre landscape through which your dog can range at will, freely and safely). My current dog, adopted as an island stray and with a tendency to be fearful of other dogs, is not a "dog park dog"; for her, long walks in quiet areas, along wilderness trails and walking paths, are a great form of exercise (and, provide her with the added mental enrichment of all the novel smells, sights, and sounds she encounters on these walks). Sometimes, a familiar dog friend or two will accompany us on these walks. The amount of daily exercise your dog requires will vary with breed and age; check with your veterinarian if you have any questions about what constitutes sufficient exercise for your own dog. And, it is important to remember that regular physical exercise with your dog benefits you, perhaps as much as it benefits your dog! Check out Suzanne Clothier's article, "Fitness in Your Own Backyard" (http://suzanneclothier.com/fitness-your-back-yard) for more tips on meeting your dog's fitness needs.

Training

Those of us who are not certified professional trainers, or who do not have much time to spare in their busy schedules, may think of training as a bit of a chore: necessary, and sometimes even

Figure 6.5: Dog sports, agility training, and so on, are great ways to provide exercise, mental enrichment, and fun for your dog. (Photos: ©Donna Kelliher, www.DonnaKelliher.com)

> **Box 6.2. Beyond the basics: types of training for enriching your dog's life, and where to go for more information**
>
> **Agility: simply great fun and exercise for you and your dog**
>
> Fenzi Dog Sports Academy (http://www.fenzidogsportsacademy.com/) is a respected online training program for a wide range of dog sports, including agility (also scent work, rally/freestyle, and even basic obedience training).
>
> Agility University (http://www.agility-u.com/) has an impressive array of online courses, ranging from basic to advanced, that you can participate in (using video of working with your dog) or just observe, as well as many other resources for agility work.
>
> Susan Garrett (http://susangarrettdogagility.com/) is a well-known Canadian trainer specializing in agility training.
>
> Greg Derrett (http://www.ultimateagility.com/) is an accomplished British trainer, whose site includes online courses.
>
> **Canine scent work**
>
> > "Inspired by working detection dogs, K9 Nose Work is the fun search and scenting activity for virtually all dogs and people." (K9 Nose Work)
>
> The National Association of Canine Nosework (http://www.k9nosework.com/) has links to certified instructors and classes, among other resources, and a link to its parent site, https://www.nacsw.net/
>
> **Flyball: this sport is fast-paced, like a relay race and high-energy game of fetch combined.**
>
> Flyball Dogs.com (https://flyballdogs.com/) contains a wealth of information about what flyball is, along with links to training, equipment suppliers, competitions, and more.
>
> The North American Flyball Association web page contains sport rules, listings of local tournaments, and much more (http://www.flyball.org/).

fun, but also very time-consuming and occasionally frustrating (particularly if we are not sure we are doing it correctly, or if successful results are not immediately obvious in our dogs). It doesn't have to be this way! We talk a lot more about training in Chapter 5, but training the basic canine life skills (like sit, stay, come when called) is best done in short bursts (even 10 minutes once a day goes a long way!), and there are numerous wonderful, rewards-based trainers and other resources out there which can help us ensure that our methods are efficient, effective, and kind to our dogs.

Importantly, however, training can be an excellent way to enrich your dog's life. In addition to training basic life skills for dogs (aka "obedience training"), there are numerous types of training beyond the basics that are fun for both you and your dog (see Box 6.2 for a few options). Over and above the mental exercise provided by these types of training, they can also provide your dog with lots of physical exercise as well – think of agility training for dogs, or Flyball. If your dog is a social butterfly, expert-led classes in these sorts of training can be a great way to increase your dog's social life, as well as providing a way for their humans to meet other like-minded, dog-savvy people. If your dog is not social, and provided you have a bit of space in your back garden, you can set up your own scaled-down agility course for your dog (Box 6.2).

Mental enrichment: more than just toys

As noted above, most of the enrichment techniques we've talked about also provide mental stimulation for your dog.

Figure 6.6: Animate enrichment: relaxed dogs at a doggy day care facility. (Photo: K.D. Setiabudi); man and his dog playing outside (Photo: iStock)

For example, an off-property walk provides your dog with a chance to investigate a world of new sights, sounds, and smells, as well as allowing her a chance to familiarize herself with the scent marks left by other dogs in her neighborhood. A training session provides your dog with a chance to learn new skills, earn rewards, and enjoy some positive interaction (i.e., fun!) with his human companion. Many of these fall under what is called in the literature "animate enrichment" (Wells 2004) – enrichment through the provision of social interaction with other dogs, or with humans. An alternate form of enrichment is known as "inanimate enrichment" – enrichment through the provision of toys, species-appropriate furnishings, and/or exposure to sights, sounds, and smells meaningful to the dogs (Wells 2004). Enrichment of either kind can increase the complexity of the dog's environment, and provide opportunities for dogs to exercise some control over their environment, two important goals of enrichment. Based on information from numerous studies, often using dogs housed in shelters, there are advantages and disadvantages to each of these two types.

Animate enrichment (contact with other dogs, and with humans) is very important for dogs. In her review of environmental enrichment for kenneled dogs, Dr. Deborah Wells (2004) describes this form of enrichment as absolutely essential. Dogs are a social species, and the majority of dogs do best when allowed to interact regularly with other members of their species. A number of studies have assessed the impacts on stress levels in shelter dogs of living in pairs (or groups) versus in solitary housing (Hetts et al. 1992; Mertens and Unshelm 1996; Beerda et al. 1999; among others). These studies found that fewer behavioral problems (such as the stress-related, repetitive behaviors) were seen, with little aggression occurring, when kenneled dogs were housed in groups versus alone. Physiological stress responses, measured by cortisol levels, were greater in dogs housed alone than in dogs housed in groups (Beerda et al. 1999), supporting the concept that enrichment enhances dogs' ability to cope with stressful situations. There are risks, of course, with housing dogs in groups; aggression between dogs may occasionally occur. For this reason, dogs should be assessed for compatibility prior to pair or group housing; indiscriminate group housing may actually be counterproductive, as outbreaks of aggression can lead to added stress or injury (Wells 2004).

Similarly, studies have repeatedly shown that even short sessions with a friendly human were beneficial in lowering stress levels (and often, improving behavior) in shelter and kennel-housed dogs. Even just two 25-minute sessions of exercise and friendly human contact (interaction in a "friendly and affectionate manner," playing, training, and walking together), two days apart, were enough to reduce stress levels and improve behavior in a study of 50 shelter-housed dogs in Spain (Menor-Campos et al. 2011). Other studies have found lowered heart rate in dogs following handling by humans (Lynch and Gantt 1968), and that dogs housed with human caretakers showed fewer fear-related behaviors than dogs born and raised in a kennels facility (Verga and Carenzi 1983). Gently handling and stroking dogs has been found to lower their stress levels in a number of studies (e.g., Hubrecht 1993; Hennessy et al. 1998).

Inanimate enrichment, in the form of toys, dog-friendly "furniture" such as comfortable beds or dens, or other sensory input (music, pheromone dispensers) can add to the complexity and interest level of your dog's environment. However, studies of the effectiveness of toys in reducing stress in kennelled dogs have yielded mixed results (Newberry 1995; Wells 2004). Some dogs interact readily and repeatedly with the toys, spending large quantities of time

Figure 6.7: Spending quality time with friendly human companions is important for your dog's quality of life. (Photos: iStock)

Mental wellness

investigating them and playing with them, and in some studies behavior improved in dogs provided regular access to toys. Other dogs largely ignore toys placed in their kennel, and provision of toys appeared to have no effect on their behavior. A few patterns emerge when looking at the literature on effectiveness of toys in enhancing quality of life for dogs: toys that make noise (with internal squeakers, for example), toys that are chewable, and soft, less robust toys tend to be more popular with kennelled dogs. One study noted that, given many dogs' preference for these sorts of toys may mean that some compromise may be needed between enrichment and safety, "since the toys preferred by dogs appear to be those that are most difficult to keep clean and pose highest risk of destruction and ingestion" (Pullen et al. 2010; p.156). For many years we shared our home with a very happy and rambunctious pit bull terrier mix named Maud, and her absolute favorite toys were the inexpensive latex squeaky toys that could be easily shredded in her strong jaws, in about 60 seconds flat (but, as we always said, it sure was a great 60 seconds for her). Your dog should be supervised while playing with toys; this is particularly vital for any type of toy that he or she can damage in this way. We always removed the shredded remains of Maud's toys before she had a chance to ingest them; we had trained a "drop it" (in exchange for a tasty treat) cue that we could employ to efficiently remove the item to a safe location. Some of the more durable toys require less supervision, as they are very difficult for the majority of dogs to damage or (assuming the size is appropriate) swallow (see Box 6.3 for some suggestions).

Another important pattern that emerges when looking at the literature on the successful use of toys as canine enrichment is that most dogs, with the possible exception of puppies (Hubrecht 1993), prefer toys that are novel. Adult dogs will habituate to the presence of a given toy quite quickly, over a day or two (DeLuca and Kranda 1992). For this reason, it is helpful to

> ### Box 6.3. Some of our favorite toys and dog-friendly devices for adding enrichment to your dog's life
>
> Kongs (https://www.kongcompany.com/): durable (red for normal chewers, black for "extreme" chewers) rubber chew toys that can be filled with soft food; recipes for filling with long-lasting treats available on the company website. Kong also makes the Kong "Wobbler", a durable toy that can be filled with dry kibble, and is easy to clean.
>
> The OmegaPaw "Tricky Treat" ball (http://www.omegapaw.com) is another option for filling with dry kibble, and is available in three sizes. Similarly, the "Toppl" from Westpaw Design (https://www.westpawdesign.com/) comes highly recommended by dog-savvy friends.
>
> Two more advanced (and, more expensive) high-tech food dispensing devices are the "Pet Tutor" (https://smartanimaltraining.com/) and the "Foobler" (https://www.youtube.com/watch?v=hEQUGy8jA44). The Pet Tutor in particular is very popular with trainers and other pet professionals, and can be activated remotely (through your smart phone, for example) for use in a wide variety of enrichment and training situations.
>
> Countless sites online offer free tips on DIY dog toys, both interactive and toys for self-play. For example, a simple, inexpensive, and durable treat dispensing toy can be made using PVC pipe (see http://www.aspcapro.org/blog/2015/03/09/tip-week-chew). Of course, anytime you are making, or purchasing, a toy for your dog, ensure that it is safe for your dog, and supervise your dog with any new toy. Ask your veterinarian or behaviorist if you have toy safety questions.

Figure 6.8: Inanimate enrichment: Maud and one of her more durable rubber chew toys. (Photo: E. Grigg)

have a variety of toys on hand, and rotate toys frequently, to maintain your dog's interest in any one particular toy.

Dogs are individuals, and thus will have preferences for certain toys, or perhaps a lack of interest in toys in general. I suspect (although I don't know that this has ever been studied) that dogs who have regular access to the outdoors and outdoor activity may be less impressed by "static" toys (toys that they play with by themselves, versus "interactive toys" that involve play with another dog or with a human) than dogs who spend the majority of their time indoors.[5] For some dogs, toys can add greatly to the quality of their environment; for others, toys may be a less effective way to add enrichment than the animate enrichment methods described above.

The wonderful world of dog toys

There are currently a truly remarkable variety of toys and devices available that are specifically designed to provide entertainment for your dog (Box 6.3 lists some of our favorites). There are also resources to help you to make your own, relatively inexpensive toys for your dog. With anything you provide to your dog, common sense precautions should be used. For example, is the toy large and sturdy enough to withstand your own dog's energy level and chewing strength, and given the size of your dog's mouth? If a commercial product, is it made by a reputable company with a history of high-quality products for dogs? In most cases, the makers of these toys will stress that your dog should be supervised when using the toy, and the toy should be removed promptly from the dog if the toy is damaged in any way (usually to avoid the dog ingesting a part of the toy, a potentially dangerous – and expensive – situation). Sadly, countless horror stories exist in the veterinary world of dogs ingesting balls that are too small for them, swallowing bits of deer antler or broken toy (and which must then be removed through expensive, invasive, and potentially risky surgeries), or consuming commercial dog chews or

---

5 I've also seen this in indoor-only cats versus indoor/outdoor cats presented with toys; indoor-only cats seem to spend more time engaging with toys.

Figure 6.9: Dogs playing with fabric snake toys at a doggy day care facility. (Photo: K.D. Setiabudi)

other food treats tainted with a toxin. If selected carefully and used wisely, however, many of these products can provide hours of fun for your dog.

One particularly popular and (in my experience) reliably engaging group of toys are those that dispense food rewards, often referred to as "food puzzle toys" or "food-dispensing toys" (Box 6.3). These toys are designed to promote foraging behavior by increasing the effort and time required to get to and consume the desired food. Ancestral dogs would have spent a significant proportion of their day seeking food, and would have expended a great deal of mental and physical energy to acquire what they needed. Today we provide them with preprepared meals, once or twice a day; these meals are usually consumed in a minute or less. And, assuming most dogs are like those that I have worked with, these precious minutes spent eating are often one of the highlights of their day. As one important goal of enrichment is to allow the dog to exercise natural behaviors, toys that allow the dogs to put some energy and work into foraging (and, increase the amount of time that dogs spend in foraging) can be very beneficial to the dog's stress and activity levels. In addition, food-dispensing toys may be meeting an emotional need for these dogs, by allowing them to express highly motivated behaviors (Morris et al. 2011). A number of studies of emotion in humans and non-human animals have described the emotional circuits of "reward acquisition" or "appetitive seeking" (e.g., Panksepp 2005; Berridge et al. 2009). These emotional circuits, it is believed, motivate many mammal and bird species to approach novel objects, and to seek out rewards (food, mates, and pleasurable experiences) in their environment. Watching the intensity and enthusiasm of a dog investigating a food puzzle toy that he knows contains a favorite food treat certainly seems to support this explanation of the value of these sorts of toys for dogs.

Food puzzle toys, it should be noted, can actually be used to feed *all* meals to your dog. I have often found this a particularly useful strategy for young and/or high-energy dogs needing additional outlets for their energy (i.e., for encouraging them to use their energy in productive ways, rather than leaving them to find other, less appealing uses for that energy, such as excessive barking or destructive behavior). Alternately, if you are in the process of training your dog and using food rewards throughout the day while doing so, at the end of the day you can put the remainder of the day's portion of kibble or other food into a food-dispensing toy. Then, allow her to work at getting the rest of her meal, adding a bit more enrichment to her day. Select food toys that can be easily and regularly cleaned. Of course, if you are consistently using food (for training and/or as enrichment in food-dispensing toys), you may need to cut down the amount of food given at regular mealtimes, to avoid unhealthy overeating in your dog. Don't hesitate to ask your veterinarian for advice on optimal weight and caloric requirements for your dog; we'll also cover this topic in Chapter 8 on physical wellness.

Other forms of enrichment to consider

Other suggested ways to enrich your dog's environment include music and other sounds, sights, and odors. For example:

- Classical music may be relaxing for many dogs. Classical music was found to increase calm behaviors in kennelled dogs; heavy metal music, on the other hand, resulted in dogs displaying more agitation (Wells et al. 2002; Kogan et al. 2012). There is even music available that is specifically designed to be soothing to dogs ("Through a Dog's Ear," available at http://throughadogsear.com/), and which has had good results in calming shelter dogs (Leeds and Wagner 2008); opinions are mixed as to whether the specially-modified music is any more effective than unaltered classical music (Kogan et al. 2012).[6]
- Many dogs feel more secure and comfortable inside a kennel or crate, and allowing dogs to access a comfortable bed inside a crate by leaving the door open for them to access at will can give them a safe retreat during the day or when things get stressful.
- In one study of kenneled dogs, platforms which allowed dogs to see beyond their own pens were heavily used by the dogs (Hubrecht 1993); for some dogs, allowing them a way to see out of a window or glass door can be a great way to add visual entertainment to their days. One exception to this would be for reactive dogs who bark excessively at passing dogs, and so on. For these dogs, it may be better to cover this view to avoid adding stress and/or perpetuating bad habits.
- Another newer alternative for adding visual and auditory stimulation to our dogs' environment could be one of the commercial "television for dogs" services. DOGTV (https://dogtv.com/) is one such channel, available online or through traditional cable services, and broadcasting programming specifically designed to appeal to dogs, particularly dogs left home

---

[6] For more on this topic, Dr. Patricia McConnell has a great article on her blog about dogs and music, available at http://www.patriciamcconnell.com/theotherendoftheleash/new-research-on-dogs-and-music

alone all day. Does DOGTV work? The website cites input from recognized and respected experts in the field of canine behavior, such as Dr. Nicholas Dodman and Victoria Stillwell, but doesn't provide any published research on the site to demonstrate effectiveness (at least, at the time of publication of this book).

- Smell is a very important sense for dogs. Although data is more limited on enrichment using smell, studies with captive wild animals have reported an increase in activity when these animals were exposed to smells such as various herbs and essential oils, musk perfumes, and scents associated with prey species (such as dung or body odors; Powell 1995; Schuett and Frase 2001; Wells and Egli 2004). Wells (2006) found that dogs exposed to lavender odor during travel spent more time resting and sitting, and less time moving and vocalizing, than untreated dogs. In my own experience, some dog owners swear by a drop of lavender on their dog's collar, or on a bandana included in their travel carrier, as an effective way to help calm them; others report seeing no effect whatsoever. It may be another example of dogs as individuals, responding differently to the same stimulus. And of course, doing canine nosework training with your dog (see Box 6.2) is a great form of enrichment allowing dogs to use their sense of smell.

- Pheromones, commercially available as Adaptil (available in collar or diffuser form, from Ceva Pharmaceuticals), have a more robust body of research behind them, with some studies reporting fewer stress-related behaviors in dogs exposed to the pheromone, in a variety of situations (e.g., Denenberg and Landsberg 2008; Kim et al. 2010). A 2010 review of the research, however, cast doubt on the validity of some of the earlier studies (Frank et al. 2010), concluding that insufficient evidence existed for the efficacy of pheromones in reducing fear and anxiety in dogs. Adaptil is often recommended for puppies and dogs of all ages, as a way to help reduce stress-related behaviors associated with encountering new experiences, unknown environments, and other stressful situations. In my own experience, it does seem to be mildly effective in reducing anxiety for some dogs, but not for others; I have not systematically studied Adaptil in companion dogs. I did co-author a pilot study on effectiveness of pheromone collars in reducing stress in dogs housed long-term in the university teaching colony; given our small sample size and the high degree of behavioral variability between the dogs, we were not able to identify any significant positive effect of pheromones on the dogs' behavior (Grigg and Piehler 2015).

Figure 6.10: One of our dogs, Elliott, relaxes after a run on the beach. (Photo: E. Grigg)

Behavioral benefits of enrichment

It is important to note that one of the most important benefits of enrichment for dog owners, beyond the beneficial effects on the dog's quality of life, are the resulting improvements in canine behavior. The primary impetus behind the studies cited above was to assess the effectiveness of enrichment, in various forms, in reducing stress-related behaviors in dogs. The vast majority of studies found strong support for the concept that, by enriching the environment of dogs, we provide them with a less stressful life, and this is reflected in their behavior.

> *Owners need to maintain not only the physical health of their pets, but their mental health as well. Providing for the mental health of pets through environmental enrichment before the development of behavior problems, may be a key factor for keeping pets in homes.*
>
> (Morris et al. 2011; p. 4235)

Take-home messages

1. Mental well-being is as important as physical health in determining quality of life in dogs.
2. Signs of chronic stress in dogs include repetitive, stereotypical behaviors (such as pacing, circling, excessive barking), and self-harming behaviors such as licking or chewing parts of the body to the point of injury.
3. Exposure to chronic stress can result in anxiety and the development of behavioral problems (or, exacerbation of existing problems). Correspondingly, the reductions in stress provided by physical and mental enrichment can help minimize anxiety and result in improved behavior.
4. Recommendations for maintaining optimal mental health in dogs focus on three aspects of their lives:
    a. environment (suitable housing, which allows dogs to meet all Five Freedoms[7]);
    b. enrichment (providing choices and opportunities to exercise natural behaviors); and
    c. daily routine (providing positive experiences each day, on as regular a schedule as possible).
5. Proven ways to provide daily enrichment for your canine companion include:
    a. social and playtime (with humans, and whenever possible, with other dogs);
    b. physical exercise (walks, games of fetch, visits to the dog park, running with you);
    c. training (in addition to the necessary "canine life skills," many options exist for training your dog in specialized skills or games);
    d. mental stimulation using puzzle toys, food-dispensing toys, and the like.
6. Always take common-sense safety precautions when providing any chews, toys, and so on, for your dog.

---

7 See Chapter 2 for a complete description of the Five Freedoms.

References and additional resources

Association of Shelter Veterinarians (ASV). (2010). S. Newbury, Ed. *Guidelines for Standards of Care in Animal Shelters*. Available online at: http://www.sheltervet.org/

Association of Zoos and Aquariums (AZA) *Enrichment*. (n.d.) Retrieved from: https://www.aza.org/enrichment/. (accessed April 2016)

Beerda, B., Schilder, M.B.H., Bernadina, W., Van Hoof, J.A.R.A.M., De Vries, H.W., & Mol, J.A. (1999). Chronic stress in dogs subjected to social and spatial restriction. II. Hormonal and Immunological Responses. *Physiology & Behavior* 66: 243–254.

Beerda, B., Schilder, M.B.H., Van Hoof, J.A.R.A.M., De Vries, H.W., & Mol, J.A. (2000). Behavioural and hormonal indicators of enduring environmental stress in dogs. *Animal Welfare* 9: 49–62.

Bekoff, M. (1972). The development of the social interaction, play, and metacommunication in mammals: An ethological perspective. *Quarterly Review of Biology* 47: 412–434.

Bekoff, M. (1977). Social communication in canids: Evidence for the evolution of a stereotyped mammalian display. *Science* 197: 1097–1099.

Bekoff, M. (1995). Play signals as punctuation: The structure of social play in canids. *Behavior* 132: 420–429.

Berridge, K.C., Robinson, T.E., & Aldridge, J.W. (2009). Dissecting components of reward: 'liking', 'wanting' and learning. *Current Opinion in Pharmacology* 9: 63–73.

DeLuca, A.M., & Kranda, K.C. (1992). Environmental enrichment in a large animal facility. *Lab Animal* 21: 38–44.

Denenberg, S., Landsberg, G.M. (2008). Effects of dog-appeasing pheromones on anxiety and fear in puppies during training and on long-term socialization. *JAVMA* 233: 1874–1882.

Donaldson, T.M., Newberry, R.C., Špinka, M., & Cloutier, S. (2002). Effects of early play experience on play behaviour of piglets after weaning. *Applied Animal Behavior Science* 79: 221–231.

Frank, D., Beauchamp, G., & Palestrini, C. (2010). Systematic review of the use of pheromones for treatment of undesirable behavior in cats and dogs. *JAVMA* 236: 1308–1316.

Grigg, E., & Piehler, M. (2015). Influence of dog appeasing pheromone (DAP) on dogs housed in a long-term kennelling facility. *Veterinary Record Open* 2: e000098.

Hennessy, M. (2013). Using hypothalamic-pituitary-adrenal measures for assessing and reducing the stress of dogs in shelters: A review. *Applied Animal Behaviour Science* 149: 1–12.

Hennessy, M.B., T. Williams, M., Miller, D.D., Douglas, C.W., & Voith, V.L. (1998). Influence of male and female petters on plasma cortisol and behaviour: Can human interaction reduce the stress of dogs in a public animal shelter? *Applied Animal Behaviour Science* 61(1): 63–77.

Hetts, S., Clark, J.D., Calpin, J.P., Arnold, C.E., & Mateo, J.M. (1992). Influence of housing conditions on beagle behaviour. *Applied Animal Behaviour Science* 34: 137–155.

Hewson, C.J., Hiby, E.G., & Bradshaw, J.W.S. (2007). Assessing quality of life in companion and kenneled dogs: A critical review. *Animal Welfare* 16: 89–95.

Hubrecht, R.C. (1993). A comparison of social and laboratory environmental enrichment methods for laboratory housed dogs. *Applied Animal Behavior Science* 37: 345–361.

Hubrecht, R.C., Serpell, J.A., & Poole, T.B. (1992). Correlates of pen size and housing conditions on the behaviour of kennelled dogs. *Applied Animal Behaviour Science* 34: 365–383.

Institute of Laboratory Animal Research (ILAR). (1996). Commission on Life Sciences, National Research Council (ILAR). *Guide for the Care and Use of Laboratory Animals*. US Department of Health and Human Service, National Institutes of Health, NIH Publication No. 86–23, 1996.

Kim, Y.M., Lee, J.K., Abd el-aty, A.M., Hwang, S.H., Lee, J.H., & Lee, S.M. (2010). Efficacy of dog-appeasing pheromone (DAP) for ameliorating separation-related behavioral signs in hospitalized dogs. *Canadian Veterinary Journal* 51(4): 380–384.

Kogan, L.R., Schoenfeld-Tacher, R., & Simon, A.A. (2012). Behavioral effects of auditory stimulation on kenneled dogs. *Journal of Veterinary Behavior* 7: 268–275.

Leeds, J., & Wagner, S. (2008). *Through a dog's ear: Using sound to improve the health & behavior of your canine companion*. Boulder, CO: Sounds True.

Lynch, J.J., & Gantt, W. (1968). The heart rate component of the social reflex in dogs: The conditional effects of petting and person. *Conditional Reflex* 3: 69–80.

Menor-Campos, D., Molleda-Carbonell, J., & Lopez-Rodriguez, R. (2011). Effects of exercise and human contact on animal welfare in a dog shelter. *Veterinary Record* 169: 388–391.

Mertens, P., & Unshelm J. (1996). Effects of group and individual housing on the behavior of kennelled dogs in animal shelters. *Anthrozoos*: 40–50.

Milgram, N.W., Head, E.A., Zicker, S.C., et al. (2004). Long term treatment with antioxidants and a program of behavioral enrichment reduces age-dependent impairment in discrimination and reversal learning in beagle dogs. *Experimental Gerontology* 39: 753–765.

Morris, C.L., Grandin, T., & Irlbeck, N.A. (2011). Companion Animals Symposium: Environmental enrichment for companion, exotic and laboratory animals. *Journal of Animal Science* 89: 4227–4238.

Newberry, R.C. (1995). Environmental enrichment: Increasing the biological relevance of captive environments. *Applied Animal Behavior Science* 44: 229–243.

Panksepp, J. (2005). Affective consciousness: Core emotional feelings in animals and humans. *Consciousness and Cognition* 14: 30–80.

Powell, D.M. (1995). Preliminary evaluation of environmental enrichment techniques for African lions (*Panthera leo*). *Animal Welfare* 4: 361–370.

Pullen, A.J., Merrill, R.J.N., & Bradshaw, J.W.S. (2010). Preference for toy types and presentations in kennel housed dogs. *Applied Animal Behavior Science* 125: 151–156.

Rooney, N.J., Gaines, S.A., & Bradshaw, J.W.S. (2007). Behavioural and glucocorticoid responses of dogs (*Canis familiaris*) to kenneling: Investigating mitigation of stress by prior habituation. *Physiology & Behavior* 92: 847–854.

Schuett, E.B, & Frase, B.A. (2001). Making scents: Using the olfactory senses for lion enrichment. *The Shape of Enrichment* 10: 1–3.

Špinka, M., Newberry, R.C., & Bekoff, M. (2001). Mammalian play: Training for the unexpected. *Quarterly Review of Biology* 76: 141–168.

Stephen, J.M., & Ledger, R.A. (2005). An audit of behavioral indicators of poor welfare in kenneled dogs in the UK. *Journal of Applied Animal Welfare Science* 8: 79–95.

Verga, M., & Carenzi, C. (1983). Behavioural tests to quantify adaption in domestic animals. In D. Smidt (Ed.), *Indicators Relevant to Farm Animal Welfare*. The Hague: Martinus Nijhoff. pp. 97–108.

Wells, D.L. (2004). A review of environmental enrichment for kennelled dogs, *Canis familiaris*. *Applied Animal Behavior Science* 85: 307–317.

Wells, D.L. (2006). Aromatherapy for travel-induced excitement in dogs. *JAVMA* 229(6): 964–967.

Wells, D.L., & Egli, J.M. (2004). The influence of olfactory enrichment on the behaviour of black-footed cats, *Felis nigripes*. *Applied Animal Behaviour Science* 85: 107–119.

Wells, D.L., Graham, L., & Hepper, P.G. (2002). The influence of auditory stimulation on the behaviour of dogs housed in a rescue shelter. *Animal Welfare* 11: 385–393.

Chapter 7

Canine behavioral problems and solutions

Photo: iStock

SOMETIMES, DESPITE ALL OUR BEST EFFORTS, WE FIND OURSELVES LIVING WITH A DOG with a behavioral problem (or two, or three …). Sometimes we acquire a new dog who comes with "baggage": behavioral issues that we, as the dog's new human family, must then find a way to modify. These issues may be due to our dog's previous life experience, in some cases exacerbated by a genetic predisposition to a certain type of behavior, such as fearfulness. At times (let's be brutally honest), we may have inadvertently contributed to the development of behavioral problems in our dogs, perhaps by not providing sufficient socialization or training for young dogs, or by allowing our dogs to spend long hours bored and alone (insufficient enrichment), or by unknowingly using punishment-based training methods which created or exacerbated fear-based behavioral issues.

Behavior problems in dogs are very common – more dogs are affected by behavior problems than any other medical condition (AAHA 2015). Behavior problems can have very serious consequences for dogs: behavior problems are a common reason for owners to request euthanasia of dogs at veterinary clinics (Landsberg et al. 2013), and are perhaps the most common reason for euthanasia for dogs under the age of 2 years (McKeown and Luescher 1988). Behavior problems are a primary reason that dogs are relinquished to shelters (Salman et al. 2000). Behavior, luckily, is not set in stone; in the vast majority of cases, behavior can be modified, using established and proven learning techniques.

When things go wrong: common behavior problems in dogs (and, what to do about them)

In this section, we will talk about some of the more common behavior problems in dogs, and what the current recommendations are (in veterinary and applied animal behavior, and based on the science of learning and behavior) for reducing or eliminating these issues. I'll start by saying that my first, and strongest, recommendation for dealing with a serious behavioral issue in your dog (such as aggression, or severe separation anxiety) is to immediately consult a professional – and not just any professional! Box 7.1 lists types of professionals trained to work with serious

Box 7.1. Getting hands-on, expert help for serious behavior problems in your dog

North America

The Animal Behavior Society certifies Applied Animal Behaviorists (CAAB) who have completed a PhD in animal behavior, completed at least three years of clinical experience, and contributed to the field via research publications. They also certify Associate CAABs (with a Master's degree, rather than a PhD). A directory of CAABs and ACAABs is available on their website (http://www.animalbehaviorsociety.org/web/applied-behavior.php).

The American College of Veterinary Behavior (ACVB) certifies veterinarians who have completed a residency treating behavior problems, and passed a challenging exam to demonstrate their knowledge and understanding of behavioral medicine. A list of diplomates can be found on their website (http://www.dacvb.org/).

The American Veterinary Society of Animal Behavior (AVSAB) has a link to "Find a Consultant," listing CAABs, DACVBs, and DVMs with special interest in behavior; (http://avsabonline.org/resources/find-consult).

UK/Europe

In the UK, the Royal Society for the Prevention of Cruelty to Animals (RSPCA) has a helpful page on locating a qualified behaviorist, including Certified Clinical Animal Behaviorists and Pet Behavior Counsellors (http://www.rspca.org.uk/adviceandadvicea/pets/general/findabehaviourist).

The European College of Animal Welfare and Behavioural Medicine (ECAWBM), like the ACVB, certifies veterinarians who have completed additional training and passed a qualifying board exam. A list of their diplomates can be found on their site (http://www.ecawbm.com/).

> **Box 7.2. Recommended general reading for more on understanding and addressing canine behavioral problems**
>
> **Books:***
>
> American College of Veterinary Behaviorists (2014). *Decoding Your Dog*. Written and edited by members of the ACVB, this is a very user-friendly book on understanding, treating, and preventing unwanted behaviors in your dog.
>
> Yin, S. (2010). *How to Behave So Your Dog Behaves* (2nd ed.) Written by the late veterinarian and behaviorist Sophia Yin, this is a lovely book packed with science-based information on canine behavior and practical tips for addressing problems.
>
> Hart, B., et al. (2006). *Canine and Feline Behavior Therapy* (2nd ed.). A textbook, but a readable one, with a wealth of scientifically-based background information on behavior problems in companion animals, as well as practical suggestions for treating problems.
>
> Hetts, S. (2014). *12 Terrible Dog Training Mistakes Owners Make That Ruin Their Dog's Behavior … And How To Avoid Them.* Dr. Hetts is a well-respected and very experienced behaviorist, and this easy-to-read book is packed full of practical suggestions.
>
> Dunbar, I. (2004). *Before & After Getting Your Puppy*. A useful, information-packed resource on puppy training and behavior, with an emphasis on avoiding future behavior problems in your dog.
>
> **Online resources**
>
> There is a *lot* of information out there on the Internet, and as we all know, not all online information is created equal. Here are a few fact-based, well-researched sites that we like, that often cover current topics in canine behavior, and can help sort the good information from the bad:
>
> The Science Dog (blog by Linda Case): https://thesciencedog.wordpress.com/
>
> Dog Spies (blog by Julie Hecht): http://blogs.scientificamerican.com/dog-spies/
>
> Eileen and Dogs (blog and informational website by Eileen Anderson): http://eileenanddogs.com/
>
> ASPCAPro (the website of the American Society for the Prevention of Cruelty to Animals, aimed at canine professionals but full of useful tips and links for all things dog): http://aspcapro.org/
>
> *Complete citations can be found in the "References and additional resources" section, at the end of this chapter.

behavioral problems in dogs (and, in many cases, cats as well). Just as it is important to carefully select the right trainer for your dog (as discussed in Chapter 5 on training), it is very important to select a qualified, experienced person to help you with your dog's behavioral issues. Anyone can call themselves a "dog behaviorist," regardless of their actual level of training or experience. Do your research, and stick with someone who has taken the time and put in the effort to get certified in this field.

There are also a number of great books out there that deal with canine behavior problems in much greater detail than we will cover here. This is an area of ongoing research, and often makes fascinating reading! Box 7.2 contains a list of some of our favorite resources for educating yourself about behavioral problems in our canine companions.

Before diving in to specific behavior problems, it makes sense to briefly review how dogs learn, as this is such an important part of understanding their behavior, and changing behavior when necessary, in effective and humane ways. We also recommend reading Chapter 5, on training, if you have not already done so – that chapter covers learning in more detail, along with specific training techniques and practical tips that provide a foundation for resolving any behavioral problem.

In addition to reviewing the principles of how dogs learn, we will introduce two additional behavioral modification techniques based on these learning principles, and which are commonly used to treat behavioral problems in dogs: systematic desensitization and counterconditioning.

Behavior modification is defined as "the intentional or structured use of conditioning or learning procedures to modify behavior."

(Hart, Hart and Bain, in *Canine and Feline Behavior Therapy*)

How dogs learn: a review of the basics

Learning processes in dogs[1] are very similar to learning processes in humans. Summarized below are the forms of learning that are most relevant to training and behavioral modification in dogs:

- *Habituation*: Habituation is an extremely simple form of learning. Essentially, after a period of prolonged or repeated exposure to a stimulus, the animal's response decreases or stops. This happens most naturally with stimuli that have no real consequences (negative, or positive) for an animal; stimuli associated with danger or negative outcomes such as pain are very difficult for animals to habituate to. The ability to habituate is a useful skill for animals for preserving energy for important life functions: animals are reacting to stimuli from the environment all the time, but they learn to "tune out" stimuli that don't provide them with useful information.
- *Associative learning*: Associative learning is simply building an association in the brain between two things (or stimuli) in the environment, or between a behavior and an outcome or response. The two forms of associative learning are classical conditioning and operant conditioning.
  - *Classical conditioning*: In classical conditioning, a previously "neutral" stimulus takes on the power to elicit a reflex-like response. Remember Pavlov's dogs (Figure 5.2 in Chapter 5): by repeatedly associating the sound of a bell with the presentation of food, Pavlov was able to build the association between the bell sound and food in the dogs' brains. After conditioning, and even in the complete absence of food, the bell sound would evoke a physical and emotional reaction in the study dogs, as they salivated in

---

1 Learning in dogs is covered in more detail in Chapter 5, on training.

anticipation of the imminent arrival of food. Classical conditioning is used in "clicker training" for dogs; the sound of the clicker device is first associated with a reward (such as a food treat), after which time it can be used to signal to the dog at the moment she has earned a reward.

- *Operant conditioning*: The other primary form of associative learning is operant conditioning, in which a certain behavior is followed by, and thus becomes associated with, a certain outcome or response. This can also be thought of as "learning based on consequences": whether or not a behavior is repeated depends largely on the consequences or outcomes for the animal the last time it performed that behavior. A behavior that is rewarded (or, "reinforced") is likely to be repeated; a behavior that is not rewarded, or is punished, is not likely to be repeated. Operant conditioning is really the foundation for dog training, simply because by controlling the outcome of a given behavior, we can control to a large degree whether or not the dog continues to perform that behavior. The process of operant conditioning is divided into four quadrants (see also Figure 5.4 in Chapter 5). Two of the four quadrants are composed of reinforcement techniques (outcomes that result in an *increase* in the likelihood and/or frequency of the behavior); the other two quadrants are composed of punishment techniques (outcomes that result in a *decrease* in likelihood and/or frequency, or cessation, of the behavior). Both techniques (reinforcement and punishment) can be either "positive" (meaning something is added in order to achieve the goal), or "negative" (meaning something is taken away):
    - *Positive reinforcement*: Something good is added (a reward of some form) in response to the behavior, in order to encourage the dog to perform that behavior again.
    - *Negative reinforcement*: Something bad (unpleasant or aversive) is taken away in response to the behavior, in order to encourage the dog to perform that behavior again.
    - *Positive punishment*: Something bad is added in response to the behavior, in order to discourage the dog from performing that behavior again.
    - *Negative punishment*: Something good is taken away in response to the behavior, in order to discourage the dog from performing that behavior again.
    - Two important points to remember when discussing the four quadrants of operant conditioning.
        1) Positive and negative do not mean "good or bad" in this usage. Instead, they are used in a more mathematical sense: *adding* something to the situation in response to behavior (positive), or *removing* something in response to behavior (negative). The term "negative reinforcement" is often misused to refer to actions which in fact fall under the category "positive punishment."
        2) From a technical (and practical) standpoint, whether or not you have actually rewarded or punished a behavior depends *not* on the consequence you imposed, but on whether or not the behavior *increased*, or *decreased*, in subsequent situations. Similarly, to be rewarding, the consequence must be of value *to the dog* (our perception of whether or not our dog should find the consequence rewarding is irrelevant). For example, for a dog who is highly food-motivated but not particularly fond of pats on the head, patting him on the head in response

to a desired behavior would not in fact be rewarding the behavior (no matter how much I think he should find my attention rewarding). Offering a food treat in response to the behavior would, for this dog, be rewarding (and thus would be more successful in increasing the frequency of the behavior).

- *Extinction*: A final learning process to be familiar with is extinction. Extinction simply refers to the fact that a response to a given stimulus can be reduced or eliminated, if the conditioned stimulus or cue is repeatedly presented without any accompanying reward or punishment. If I have a dog who begs for food scraps at the table, one way to reduce this behavior is to completely ignore this behavior from the dog. If the behavior (begging at the table) is *never* rewarded, eventually the dog will stop performing the behavior. Note that this may not be the fastest way to eliminate begging at the table, but if done consistently, it will work (eventually). Extinction as a strategy for reducing behavior only works, of course, if we control the reward. I can ignore my dog digging holes in my flower beds until the cows come home, but that won't stop the behavior. The enjoyment for my dog comes from the activity itself, and she doesn't need my participation for it to be rewarding! And, there is a flipside to this coin, of course. If I have trained a desirable behavior (such as "come when called") by initially rewarding the dog for coming each time I call her, but then (once she has learned the behavior) I simply stop all rewards and *never* reward her for coming when called, I risk a decline in her performance over time. Rather than completely eliminating rewards for successful performance, it is better to reduce the frequency of rewards (so, once she has clearly demonstrated that she knows the cue and behavior, she does not earn a reward *every* time she completes it successfully, but instead is rewarded only periodically). Based on my experience, from a practical standpoint, I believe in being willing to reward our dogs periodically for complying with our requests. Doing so provides continuing motivation and promotes durability of performance, and things like a very reliable recall can be very useful in those moments when you really, really need your dog to comply with your requests.

Systematic desensitization

As noted earlier, animals have the ability to habituate to repeated, harmless stimuli. Habituation, again, is evidenced by the animal's reaction to the stimulus decreasing or stopping completely. There may be times, in raising and living with our dogs, when we would really like them to stop reacting to some stimulus in their environment, particularly when their reactions are negative ones: barking, lunging, or growling. Systematic desensitization is the process of setting the dog up to habituate to a particular stimulus. The way the process works is simple, although it can require patience.

> *Behavior is not necessarily reasonable. If it has already been established as a way to earn reinforcement, and if the motivation and circumstances that elicit the behavior are present, the behavior is likely to manifest itself again.*
>
> (Karen Pryor, in *Don't Shoot the Dog! The new art of teaching and training*)

Whatever stimulus the dog is reacting to, the first steps are to determine exactly what the

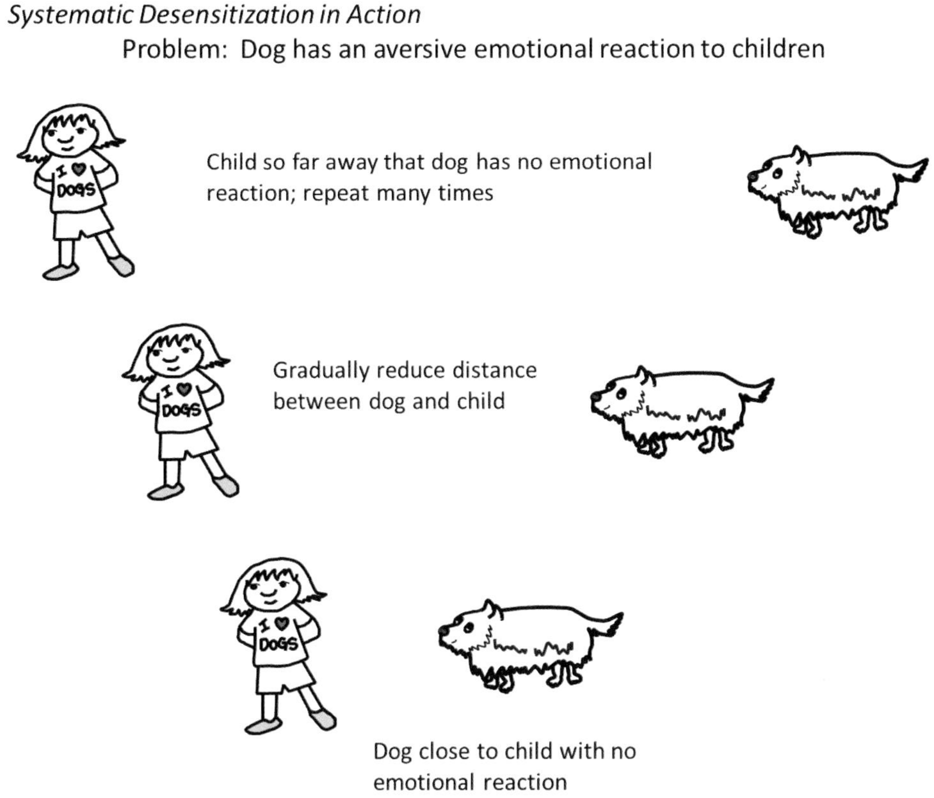

Systematic Desensitization in Action
    Problem: Dog has an aversive emotional reaction to children

 Child so far away that dog has no emotional reaction; repeat many times

 Gradually reduce distance between dog and child

Dog close to child with no emotional reaction

Figure 7.1: Systematic desensitization. (Adapted from a diagram presented by J. Neilson, DVM, DACVB, at Ross University, 2011)

stimulus is, and then to identify a gradient or series of increasing levels of intensity for this particular stimulus. So, for example, if the dog is reacting to the presence of unfamiliar people by growling or barking, the level of the stimulus (an unfamiliar person) might be changeable by varying the distance between the dog and the stimulus. The unfamiliar person located very far away from (but within view of) the dog would represent a very low level of stimulus intensity (as evidenced by the dog noticing, but not reacting in any way to, the stranger). On the other end of the gradient, the unfamiliar person within a meter or two of the dog would represent a very high level of stimulus intensity (by this point, this dog is likely "over threshold" – i.e., reacting strongly to the proximity of the scary stranger). What constitutes "very far away" will vary dog to dog; part of these first steps is to identify what constitutes a safe level (sufficient distance, in this case) for your dog.

The next step is to begin exposing the dog to the stimulus, *at the lowest possible level* (i.e., a level at which the dog is not reacting negatively in any way; Figure 7.1). Work at this level for a few sessions – practicing easy cues like "watch me," "sit," or similar, and rewarding the dog for responding appropriately. After a few sessions at this level, begin to slowly and gradually increase the level of stimulus for the dog. The key here is to go slowly; this is where patience is required! If the dog begins to react negatively to the stimulus, you have tried to go too fast (a very common mistake); go back to the previous level, and increase by a smaller increment next time. The goal is to keep the dog "under threshold" (i.e., not reacting to the stimulus, and able to remain calm), while gradually increasing the intensity of the trigger. Keep these exposure sessions short: sessions of 10 or 15 minutes at a time, once or twice a day (and every day, if you can manage), are ideal. Do not move to a higher intensity of stimulus until you have had a few sessions at the previous level during which the dog showed no negative reaction to the stimulus. You're looking for the dog to ignore the stimulus, or show at most mild interest, before moving to the next level. Over time, you will find that the dog tolerates higher and higher levels of the stimulus, with the goal being habituation: at some point, the dog will no longer display any negative reaction to the stimulus. If you are lucky, you have noticed your dog's reactivity to the stimulus early (when your dog is young), and are working to reduce the behavior promptly; the process of systematic desensitization can go quite quickly with early intervention. If not, it may take quite some time to get the dog to habituate to the stimulus. In general, the longer a dog has been performing these behaviors (and the stronger the dog's reaction to the stimulus), the longer it takes to reduce them. Practice makes perfect, after all, and a dog who has been behaving in a certain way for years has had a long time to perfect the behavior. But, don't give up! I have seen this technique work in many dogs, even dogs with a long history of reactivity; be patient, be persistent, and have faith that your dog's behavior will improve – and it will.

Thankfully, there is another technique, commonly used in conjunction with systematic desensitization, which can significantly improve the speed and effectiveness of that process (Figure 7.2). It's called counterconditioning, and we'll talk about that next.

Be patient, persistent, and have faith that your dog's behavior will improve ... and it will!

Counterconditioning

Counterconditioning is a form of associative learning, and a variant of classical conditioning. Associative learning, as defined earlier, is the process of building an association in the dog's brain between two things (or stimuli) in the environment, or between a behavior and an outcome or response. In the case of counterconditioning, the goal of the conditioning process is to establish a *new, different* emotional response to a given stimulus; in other words, to *replace* an existing emotional response (usually an undesirable one, such as fear or aggression) with one that is incompatible with that response, and more desirable (such as positive anticipation of a reward or pleasurable event).

Counterconditioning is accomplished simply by repeatedly and consistently pairing the stimulus with a pleasurable outcome for the dog. In practice, with food-motivated dogs, the easiest way to do this is by pairing the sight (sound, smell, etc.) of the stimulus with a small, high-value food reward. The stimulus should, in effect, come to be a predictor of good things for the dog, and from the dog's perspective. Going back to our example of the dog who barks and lunges at strangers, every time the dog sees a stranger, he should get a small reward; ideally, the dog is rewarded when he is still under threshold, and behaving calmly and appropriately. Working with this dog will require having rewards (food treats, or a favorite toy) readily available and on hand anytime he is likely to encounter an unfamiliar person, until he begins to associate

Systematic Desensitization with Counterconditioning:
Problem: Dog has an aversive emotional reaction to children

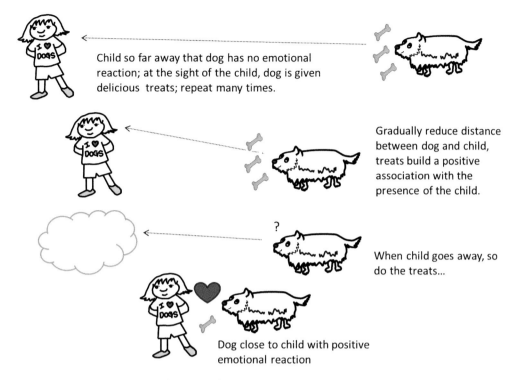

Figure 7.2: Systematic desensitization with counterconditioning. (Adapted from a diagram presented by J. Neilson, DVM, DACVB, at Ross University, 2011)

the presence of strangers with these rewards. Keeping him under threshold will involve some management of encounters early in the process, which is why the combined techniques of systematic desensitization and counterconditioning work so effectively when paired. As one behaviorist I know phrased it, "For the dog, the sight of a stranger should come to mean, 'Treats will rain from the sky!'".

A note about "flooding": Flooding is a behavioral therapy technique used to treat fears and phobias in humans, in which the person with the fears is exposed to the thing that frightens them for a sustained period of time.[2] At times, it may be tempting to think that, if we just expose the dog to whatever scares her at high intensity and for a sustained period of time, she will eventually just get used to it. And, wouldn't this be faster than the gradual process of systematic desensitization and counterconditioning? There are two primary objections to taking this approach. The first is that, in many cases, this approach can make the situation worse. When "flooded" by the thing that scares her, the dog can become overwhelmed, and is likely to have a very bad emotional experience in the presence of that stimulus (even if she is not physically injured in any

---

[2] Although this technique, also called "exposure therapy," is still used in human psychotherapy today, it is regarded by many therapists as more stressful than systematic desensitization; in addition, human patients are generally taught self-relaxation techniques prior to exposure to the frightening stimulus.

way during the encounter; Hart et al. 2006). This may simply *reinforce* her negative associations with the stimulus, rather than reducing them, making her even more sensitive to this stimulus. And, if the fear reaches the point where the dog feels the need to defend herself, it may put other people, animals, and so on, in the vicinity at risk from the frightened dog. The second is a question of ethics: is it ethical to put our dogs into a situation that we know will upset or terrify them, when there is another, proven, kinder way to change their reactions to a given stimulus?

Of course, in all of this, don't forget the basics of rewards-based dog training, and *operant conditioning!* Always remember, a response followed by reward ("reinforcer") will increase the probability of response occurring again. Dogs, like humans, will tend to repeat an action that they have found rewarding – in other words, dogs go with what works! So, when faced with a behavior that you are not comfortable with, decide what you would like your dog to do *instead*, and then teach/reinforce that behavior! Review the training techniques and tips in Chapter 5 to maximize your training success.

The prelude to addressing behavioral problems – be sure it really is "just behavioral"!

Before assuming that the problem you are having with your dog is purely behavior, it is important to look for and rule out (or, treat as necessary) any physical or medical issues that may be contributing to the behavior. This is a "golden rule" in veterinary behavioral practice, as changes in behavior often are a sign (and often the first sign) of an underlying medical problem (Box 7.3).

Pain, for example, is a common factor that can trigger or exacerbate aggressive behavior in dogs. Pain can be felt anywhere in the body, and can be acute (sudden onset, often sharp pain) or chronic (persistent, long-lasting pain). Causes may include injury, trauma, illness, surgical procedures, or joint disease such osteoarthritis (common in older animals). Dogs are experts at masking pain, and it is important to remember that pain may be present before the dog begins to limp, or display obvious lameness. Common signs of pain are often behavioral: vocalizations, agitation, abnormal gait or posture, a loss of appetite, and/or increased aggression.

Humans aren't the only animals to suffer from emotional thunderstorms that make our lives more difficult, and sometimes impossible ... Every animal with a mind has the capacity to lose hold of it from time to time."

(Laurel Braitman, (2014) *Animal Madness*)

> **Box 7.3. Common behavioral signs of an underlying physical health issue in dogs**
>
> Change in eating habits (resulting in marked weight loss or gain).
>
> Change in elimination habits (for example, loss of housetraining, inability to control urination).
>
> Change in activity levels (dog is notably more, or less, active than in the past).

If you are having a serious behavioral problem with your dog, particularly if your dog's behavior has changed suddenly or inexplicably, always work with your veterinarian to rule out any underlying medical problems.

In some cases, you may need to be persistent when investigating possible underlying medical problems. And, pain (whether clearly present or highly probable) should always be considered and treated. Pain measurement scales (such as the Helsinki Chronic Pain Index, or the Glasgow Composite Measures Pain Scale) exist that can be helpful in identifying existence and level of pain in your dog; your veterinarian should be able to provide you with one or other of these forms (or complete one for you). In the past, pain in companion animals has tended to be dismissed or undertreated; this is no doubt a legacy of earlier beliefs that non-human animals did not actually feel pain. Thankfully, the science of the management of pain in companion animals has improved markedly in recent years (AAHA/AAFP 2015), but don't hesitate to ask your veterinarian about treating pain in your canine companions, whenever you have concerns. Do not, however, administer *any* pain medications to your dog without first checking with your veterinarian; some pain medications used in humans are toxic to dogs, and even for drugs safe for use in most dogs, it is important to know the correct, safe dosage for your dog!

When things go wrong, continued … four common behavior problems in dogs

The following section consists of an overview of four categories of behavioral problem:

- **"Unruly," "rude," or "nuisance" behaviors:** These are the relatively mild (but often very frustrating) behaviors like jumping up, excessive barking, and boisterous, overly-energetic play with humans and/or other dogs. These behaviors can often be reduced just through revisiting basic "life skills" training for dogs, combined with a few other techniques and tips.
- **Aggression:** This is by far the most common, and most serious, behavioral complaint that we deal with as professional behaviorists. We'll talk about possible causes, required safety measures, and techniques for reducing aggressive behaviors.
- **Fears, phobias, and anxiety:** This is another very common area of behavioral problems for dogs, and an important one to address early, as fearfulness can often transition into aggression later in life. We'll talk about why this happens, along with common causes of fear in dogs, and methods for reducing fear and increasing calm and self-confidence in dogs.
- **Separation anxiety (SA):** This is a sub-category of behavioral problems involving anxiety, but a particularly frustrating one for owners as the problems occur when they are not present with the dog to address these behaviors. We'll talk about what is currently known about SA in dogs, and some techniques and approaches for reducing these behaviors.

"Nuisance behaviors" or "normal behaviors"? Either way, cute in puppies, less so in full-grown adult dogs

In many cases, the category of behavioral problems often classified as "nuisance behaviors" are in fact perfectly normal behaviors, from a dog's standpoint. The majority of dogs bark, many like to play roughly, puppies chew things and are often mouthy as they go through their teething phase, most dogs will happily get up on the comfortable sofa and bed if we let them, and many will jump up on us when excited (particularly if these behaviors work for them, in getting them something they want, including our attention). In some cases, we allow these behaviors and even reward them when our dogs are puppies; then, when the dog is full-grown, we realize that these behaviors are no longer pleasant or even safe, and we try to change them. For example, many dogs will happily and excitedly jump up on us when greeting, particularly if we have been separated from the dog for a while. Most of us love it when our dogs are ecstatically happy to see us on our return home; for many of us, this is one of the most rewarding aspects of sharing our home with a dog! When dogs are young and small, this jumping up doesn't pose much of a problem. We initially reward the dog when she jumps up, focusing our attention on her, speaking to her in a happy voice and patting her head. As the dog grows larger, however, this behavior can become more uncomfortable, and even risky, particularly for children and senior citizens. This behavior, unfortunately, can be very persistent, as it tends to occur when the dog is excited, and may have been consistently rewarded in the past. So, where do we go from here?

We don't really want to discourage our dogs from being pleased to see us, so using positive punishment methods (by yelling, stepping on their toes, smacking them on the nose, or kneeing them in the chest) doesn't seem like the best approach. In addition, in many cases these types of techniques, although they may end the behavior in the moment, will not stop the behavior from reoccurring in the future. If the dog is seeking attention (a primary motivator for dogs jumping up on people), and still sees the yelling, and so on, as attention from and interaction with the owner, then from a functional standpoint, the behavior continues to be rewarded. Behavior that is rewarded, given basic learning principles, will continue. If instead we take a step back (metaphorically and physically), and think about:

1. why the dog is doing the behavior (what is their goal, and what is rewarding the behavior); and
2. what alternate behavior we would rather the dog did in that situation; and
3. can we manage the dog's environment to minimize the chance that the behavior will happen;

then we can begin to put together a logical, effective plan for changing the behavior. For example:

1. Your dog may be jumping up to get your attention (and, any attention you give to your dog in this moment is rewarding this behavior). Can you remove your attention in response to the jumping up, instead? For example, only greet your dog when he is calm. If he jumps up on you, immediately and pointedly turn your back on him, looking away and not interacting with him in any way. If he continues to jump on you even when your back is turned, you may need to leave the room, step over a baby gate into another room, or similar. The moment

> *Training occurs every time you interact with your dog. Be aware of your every action and, until your dog's good behavior is a habit, be ready to reward desirable behavior as it occurs and remove rewards for undesirable behaviors before they are rewarded.*
>
> (Sophia Yin (2010), *How to Behave So Your Dog Behaves*)

he stops jumping up, and either has all four feet on the floor or sits down, turn and greet him. The goal is to communicate to your dog that jumping on you results in an immediate removal of your attention; calm behavior, on the other hand, will be rewarded with your attention and praise. It is important to ensure that others interacting with your dog follow the same set of rules that you do; otherwise you may end up with a dog that does not jump up on you, but will still jump up on others who continue to reward this behavior.

2. Perhaps you would rather your dog sat calmly on his bed or mat (or, simply sat calmly) when you arrive home at night, and you could then greet a calm, (relatively) stationary dog. In this case, train the behavior ("sit," "stay," "place," or "go to bed," etc.) using rewards-based training, and generously reward him the first time he successfully stays sitting as you enter the house (rather than jumping up). This approach, of training an alternate (often incompatible) behavior in a given situation or in response to a specific trigger in the environment, is often called "*response substitution.*" Dogs look to us for guidance – give your dog another, more appropriate way to succeed (to get what he wants or needs) in the situation. You may find it most effective to combine both approaches: training and rewarding the preferred behavior, and removing your attention (negative punishment) for mistakes or relapses.

3. To make it more difficult for your dog to continue this behavior, you might try installing a baby (or, pet) barrier or gate, temporarily, between the door and the rest of the room or house. Your dog must remain on the far side of this barrier as people enter the house. Once he has a chance to get a look at the new person, and sits voluntarily (or at first, at least responds correctly to the cue for sit), the new person can move through the gate to say hello. Others may find that enclosing the dog in her crate or behind a closed door while guests are entering, and then allowing the dog to come out and greet guests once the guests are settled, is helpful. For many behavioral issues, managing the dog's environment to reduce the opportunities that he has to practice the undesired behavior (or at least, reducing the opportunities to do so when you are not present to intervene), is an important part of successfully changing the behavior.

Other similar issues, such as barking, roughhousing, and so on, can be approached in a similar way. Think about why the dog is doing the behavior, how it is working for the dog (i.e., what is rewarding the behavior, causing it to continue), and what else you would rather the dog was doing in that situation. Then, set up your training plan. Be aware that, particularly for things like barking, other issues may be causing or exacerbating the behavior. You may need to do other types of behavioral modification work to reduce these behaviors. For example, if your dog explodes into raucous territorial barking at the sound of the doorbell, you may need to do some systematic desensitization and counterconditioning, to help her habituate to the sound of the doorbell, and/or to see the doorbell as a predictor of good things (e.g., if she remains calm, doorbells signal a small food treat and the chance to greet a new person). She may need to learn alternate responses to the sound that will earn her a reward (for example, checking in and alerting you to the presence of someone at the door, by nudging your hand or similar; going to fetch a favorite toy; or immediately going to sit on her bed near the door).

Don't forget during this process that, just as it is important not to inadvertently reward the undesirable behavior, it is equally important to reward the alternate, desirable behavior. I frequently see owners successfully removing the rewards for the behavior as a way to reduce the behavior (for example, turning away when the dog jumps up on them), but sometimes they then forget to reward the dog for the correct behavior (sitting calmly and waiting to be greeted). Remember, the goal is to teach the dog another way to greet people that will still earn her the reward she seeks.

One of the best ways to avoid ever seeing these sorts of problems is to work on teaching your new addition self-control, rewarding calm and appropriate behaviors, and ensuring he is well socialized with people and other animals, from the point when he first joins your household. There are a number of great books on raising a puppy that provide step-by-step tips for achieving this goal; see Chapter 5 on training for a list of recommended resources (as well as more information on the process and importance of socialization). I often encourage my students to teach their puppies "sit to say please": any time the dog wants something, he sits to "ask" for it (rather than barking, whining, pawing, etc.) – whether it is attention, a treat, or so on. A similar (although stricter) approach is sometimes called "nothing in life is free"[3] (except water and air) or "learn to earn" (S. Yin's program[4]), in which dogs need to "work" – often meaning doing something as simple as a "sit" or "down" – to get the good things in life. So, for example, before your dog is allowed out in the back garden for playtime, you ask him to sit at the door; the moment he does, you praise him and open the door for him. Same with serving him meals, giving him chew toys, and so on. In my experience, consistently using this approach encourages self-control (or, impulse control) in dogs, giving them calm, controlled ways to get the things they want. On a final note, there may be times when it is worth just stepping back and enjoying the essential "dogness" of your dog. Don't punish your dog for being a dog (but you can, and in some cases should, train alternatives to the less appealing doggy traits).

Aggression

Although it is a few years old now, I still think one of the best articles on canine aggression towards humans is Dr. Ilona Reisner's "Differential diagnosis and management of human-directed aggression in dogs" (2003). The article is written for a veterinary audience, but in it Dr. Reisner clearly outlines the most common reasons for canine aggression, important safety and management issues, and recommended behavioral modification techniques. In it, she makes four very important points about aggression:

1. Each aggression case is unique (because each dog, family, environment, etc., is unique).
2. Aggression cannot be "cured" (aggression, in many cases, is a normal canine behavior, although

---

3 I don't know who first coined the term "nothing in life is free"; I first heard of it from the veterinary behaviorist and educator Dr. Lori Gaskins. Dr. Ian Dunbar's website, "DogStarDaily," has two relevant blog posts discussing the advantages and potential disadvantages of this technique, at http://www.dogstardaily.com/blogs/nothing-life-free (on the basics) and http://www.dogstardaily.com/blogs/nilif-nasty (on the possible implications of too strictly adhering to this approach).
4 More on "learn to earn" can be found in Dr. Yin's 2010 book, *How to Behave So Your Dog Behaves*, pp 118–123.

the level of aggression seen may be abnormal; aggression must be managed throughout the life of the dog for safety of all concerned!).
3. A clear, feasible (given the people in the household, the household schedules and other logistics, etc.) plan for how to keep everyone safe *must be* the *first* step.
4. Owners should realize the risk is lifelong (aggressive behavior can be modified and reduced in almost all cases, but this shouldn't mean we should become complacent and put the dog back into situations where aggression predictably occurred in the past, without taking appropriate precautions).

Aggression in dogs can occur for a number of reasons, including fear-based or defensive aggression, dog-to-dog aggression, predatory aggression (usually towards small mammals such as cats, although on rare occasions may be directed to very small children), territorial aggression, owner-directed aggression (such as resource guarding, or food-related aggression), redirected aggression (aggression towards an external trigger that is redirected onto a nearby dog or human), and idiopathic aggression (quite simply, aggression that happens for reasons that we don't understand, but may be linked to an underlying medical or neurological issue). Just because a dog displays one form of aggression, does not necessarily mean that he will display any of the other forms. Aggression (other than perhaps idiopathic aggression) is almost always in response to a very specific trigger or set of triggers; if the trigger is not present, the aggression is unlikely to be seen. The trick often is, to be able to identify what the trigger is, and to accurately read the body language of our dogs, to know when aggression is imminent (and, intervening *before* the aggression happens). We have all heard someone say, following an aggressive incident, that the aggression "came out of nowhere" and the dog "gave no warning." Sadly, in the vast majority of cases, this is not really true – we have missed the trigger, and/or been too slow (or just unable) to read our dog's increasing level of fear or tension, and thus we did not intervene quickly enough to prevent the incident. These incidents can have tragic outcomes, both for the humans and dogs involved, and we owe it to them and to ourselves, to learn everything we can about how to keep everyone safe.

For any dog, we can minimize the chances of ever seeing aggression by ensuring that young dogs get sufficient and appropriately-conducted socialization early in their lives (we talk more about the process of socialization in Chapter 5, on training). As Dr. Ian Dunbar has said, "Dogs should meet 100 new people in the first three months of life" – especially children! Meetings should always be supervised and accompanied by good things for the dog – small food rewards are an easy way to build up a positive association in the dog's mind with new people, children, and so on. Remember that dogs often look to us for guidance in how to respond in a given situation; keep your body language and tone of voice calm and happy when greeting new people and new situations. One reason why the importance of socialization is so well established is that, for dogs, unchecked fear often

> *Aggression is perhaps the most challenging canine behavioral problem presented to behavioral specialists and continues to be the most common reason for referral to veterinary behaviorists (Beaver 1994), at least in part because of the emotional stress and risks of living with a biting dog.*
>
> (Ilona Reisner, 2003, p. 303)

Box 7.4. Additional resources for understanding and working with canine aggression*

Books

Parsons, E. (2005) *Click to Calm: Healing the aggressive dog*. Published by Karen Pryor's renowned training organization, this is a wonderful how-to manual on using gentler, rewards-based training to reduce aggression and promote calmer, more appropriate behavior.

McConnell, P. and London, K. (2009) *Feisty Fido: Help for the leash-reactive dog*. Specifically aimed at dogs who are reactive to other dogs on-leash, and full of step-by-step instructions for changing this behavior.

Aloff, B. (2002) *Aggression in Dogs: Practical management, prevention and behaviour modification*. This is a lengthy book, covering all aspects of canine aggression, including in-depth information on to how to reduce it.

Stewart, G. (2012) *Behavior Adjustment Training: BAT for fear, frustration and aggression in dogs*. While not all behaviorists subscribe to BAT, many owners and trainers do find the techniques helpful in redirecting and reshaping reactive behaviors. Worth a read.

Donaldson, J. (2004) *Fight! A Practical Guide to the Treatment of Dog–Dog Aggression*. Specifically focused on understanding and reducing dog aggression towards other dogs, from the well-respected trainer Jean Donaldson.

DVDs

McConnell, P. (2010) *Treating Dog-Dog Reactivity*. Dr. McConnell is an accomplished public speaker, and this is a fascinating series of lectures on this topic, complete with many helpful suggestions and technique tips.

Yin, S. (2014) *Dog Aggression: From fearful, reactive and hyperactive to focused, happy and calm*. Dr. Yin presents a comprehensive plan for behavioral modification to reduce reactivity, and so on, and increase calm in your dog.

McConnell, P. (2004) *Reading Between the Lines*. This DVD is specifically focused on improving our ability to read body language in our dogs; ability to accurately "read" your dog is a prerequisite to working with aggression in any form.

Websites

The CARE program (http://careforreactivedogs.com/) for reactive dogs, from trainer and behavioral consultant Jennifer Titus, is a step-by-step program (based on learning theory and using rewards-based methods) for reducing reactive behaviors like barking, lunging, and pulling on leash.

*Complete citations are available in the "References and additional resources" section, at the end of this chapter.

turns into aggression later in life. Initially, a dog who is fearful will usually do all it can to escape the scary situation: backing away, cringing, turning away, and/or leaving the area (if he can – dogs on-leash are not at liberty to move away freely). If the scary situation continues (the scary person continues to approach, the other dog continues to glare challengingly at him, or similar), the dog may eventually reach the point where he feels he has no option other than to defend himself. The problem is that, once pushed to the point of defensive aggression (growling, lunging, biting), the dog may quickly learn that this aggression is a highly effective way of making the scary situation less scary (the scary person backs off, the owner of the glaring dog – finally – removes her to a safe distance, and so on). Once they learn this lesson, the next time they are faced with a similar situation, they may not go back to square one, instead choosing to go with the behavior that worked so well the first time – aggression. In many cases, the more often that they repeat this behavior successfully, the less the behavior looks like fear-based aggression, and the more it evolves to look like confident, assertive aggression. We may wind up with a dog who does not give us the warning signals usually displayed by a fearful dog, and instead goes straight to the strategy that works. It is worth reviewing Figure 4.3 (in Chapter 4, on body language), showing the Canine Ladder of Aggression, which depicts the progression and changes in canine body posture from relaxed to tense to aggressive.

I would reiterate again here that, if you are experiencing a serious aggression issue with your dog, the best course of action is to a) speak to your veterinarian (primarily to rule out any underlying medical issues that may be causing or contributing to the aggression), and b) seek help promptly from a qualified professional in canine behavior (see Box 7.1 for where to find hands-on help). There are tried-and-true techniques out there to help reduce and manage aggressive behavior, and some wonderful professionals with years of successful experience in helping dogs and their owners through this stressful time. Don't hesitate to get the help you need (and, seek help sooner rather than later – behavior problems are much easier to treat when caught early on).

So, what follows are some baseline suggestions for how to manage when faced with aggressive behavior in your dog (but, these suggestions are not meant to substitute for working with a behaviorist, in more serious cases of aggression). Box 7.4 lists some additional recommended resources for working with various forms of aggression.

First steps for any type of aggression (including what to do in the moment, when faced with aggressive behavior)

1. As discussed earlier, a full medical exam for your dog is the first step, to ensure that there isn't an underlying medical reason (such as pain) fueling this behavior. If your dog is an intact male, you may also want to consider neutering your dog; a 1997 study by Dr. Jacqui Neilson and colleagues found that castration resulted in marked reductions in aggressive behavior towards humans and unfamiliar dogs in many (although not all) male dogs (Neilson et al. 1997). As there is a genetic component to most behavior, breeding a highly aggressive dog is not recommended.
2. The next step is to have a realistic, effective safety plan in place for living and interacting with the dog, while you are working on reducing the aggressive behavior. This might involve having the dog wear a basket muzzle and/or dragging a leash in all situations in which he may become aggressive (for some dogs, this would extend to having the dog muzzled any time he is outside the home or in the presence of strangers). Muzzles have gotten a bad rap in the past, and accompanying a dog wearing a muzzle can be a stressful experience, as many people

tend to view dogs wearing muzzles as "bad dogs" (and often extend this condemnation to their owners). This is unfair (and, counterproductive, if it discourages people using muzzles on dogs that really should wear one, for safety's sake). A muzzle is a safety tool, and wearing a muzzle makes the dog, and everyone interacting with the dog, much, much safer during behavioral modification work, necessary physical exercise, and social opportunities for the dog. Putting a muzzle on your dog, in that sense, indicates your dedication both to the welfare of your dog, and the safety of others! And, numerous resources exist for how to introduce your dog to a muzzle so that your dog is not fearful of wearing one, and in fact comes to see the muzzle as the gateway to food rewards, outdoor fun, and so on.[5] Introducing your dog to the muzzle in an appropriate way (i.e., using lots of high-value food rewards and/or other favorite activities to build up a positive association with the muzzle) is an important first step in use of this tool. If not introduced carefully, most dogs will (understandably, perhaps) hate wearing muzzles when they are first put on. Use a basket muzzle, not a nylon one; these are safer for the dog (dogs can breathe more easily) and they can be fed food rewards for appropriate behavior through the openings. Be sure, too, that you get a muzzle that fits your dog correctly; most companies that manufacture or distribute muzzles can provide detailed instructions on how to measure your dog for correct fit. The Muzzle Up! Project (https://muzzleupproject.com/) is a great resource for dog owners considering using a muzzle for their dog, as well as those working with their dog on a muzzle. The Muzzle Up! Project was started by trainer Maureen Backman to change the stigma trainers and dog owners face when they put a muzzle on their dog.

3. Other safety recommendations for living with an aggressive dog include (adapted from Horwitz and Neilson 2007):

   a. **All warnings from the dog should be heeded.**

   In other words, be sure that you (and any others who interact with the dog) understand canine communication, in the form of body language and vocalizations (body becomes stiff or tense, growling, lift lip, etc. – see Chapter 4 on canine body language for more details). And, importantly, respect these warnings! It is not worth the risks to challenge a dog who is behaving aggressively. Better to back off – get everyone to a safe distance, allow the dog to calm down, and then get on with the management and behavioral modification plan to reduce the aggressive behaviors. In many cases, aggression occurs because the dog is trying to guard some resource he considers valuable; in these instances, it may be possible to distract and redirect the dog out of the situation (shake the treat jar, toss a high-value toy or treat to a safe distance), shifting him away from the resource (and, shifting his behavior away from aggression).

   b. **No interaction when animal is aggressive.**

   As above, do not risk challenging a dog who is behaving aggressively. Alternately, don't try to reassure him as a way of placating him out of the aggression (you don't want to risk

---

[5] Three good online videos available free on YouTube and demonstrating the process of desensitizing your dog to wearing a muzzle are: Dr. Lori Gaskin's video at https://www.youtube.com/watch?v=IDccJQED-MM; Kikopup's video 'Focusing on Calmness' at https://www.youtube.com/watch?v=KJTucFnmAbw; and Domestic Manners dog training's video at https://www.youtube.com/watch?v=1FABgZTFvHo

inadvertently reinforcing the aggressive behavior). Instead, immediately and pointedly move away from the dog (in a sense, this is negative punishment for the behavior, as you are removing your attention from the dog). And, as soon as the dog is calm, begin work to change your dog's reactions to the situation, so that the aggressive behavior is reduced or eliminated.

c. **Avoid situations and triggers that provoke aggression.**

Although serious canine aggression is, of course, not a laughing matter, this recommendation always reminds me a bit of the old joke involving a man consulting his doctor for help with pain in his arm. "Doctor," says the patient, raising his arm over his head, "It hurts when I raise my arm like this!" "So," replies the doctor, "Don't raise your arm like that!" Part of successfully managing and modifying aggressive behavior is to do everything possible to avoid putting the dog into situations that you know are likely to cause her to behave aggressively. So, for example:

1. If the dog is aggressive while eating, leave the dog alone while eating, in possession of chews or food treats, and so on.
2. If the dog is aggressive towards children, *never* leave children alone with the dog.
3. If the dog is aggressive towards visitors, confine the dog before visitors enter, and take measures necessary to ensure the safety of visitors around the dog.

d. **Avoid physical reprimands (positive punishment).**

Studies have shown that "confrontational techniques" (positive punishment approaches like "alpha rolls" or forcing the dog onto his back, yelling at the dog, hitting the dog, etc.) actually put owners at a higher risk of increased aggression from their dog (Herron et al. 2009). This is particularly true if the dog has ever shown any signs of owner-directed aggression. In that study, *other approaches to changing the dog's behavior to something safer and more acceptable proved highly effective and much less risky.* These approaches included use of food rewards to train alternate behaviors (87 percent of 140 dog owners who had brought their dogs in for behavioral consultations for aggression reported that this approach had worked well for them), teaching "Sit to say please," aka "sit for everything" (85 percent of owners reported improvements), using food to trade for item (86 percent of owners reported improvements), and clicker training (65 percent of owners reported improvements). Only 2 percent of owners reported that these approaches had made behaviors worse, in contrast to 43 percent of owners reporting that hitting or kicking their dog made the dog's behavior worse (not surprisingly, when you consider that this treatment would have almost certainly caused defensive aggression in the dog). Similarly, 31 percent of owners who had tried the "alpha roll" reported that this made the problem worse rather than better, 41 percent of owners who had growled at their dog reported that this made the behavior worse, and so on (Herron et al. 2009).

The next steps: treating aggressive behavior in dogs

As Dr. Reisner noted (2003), each canine aggression case is unique. Which of the recommendations listed below will be most useful for you will depend on a number of factors, such as the type of aggression you are seeing, the nature of your situation and household, and so on. The following guidelines are adapted from a set of recommendations for veterinarians helping owners with aggression in their canine clients (Horwitz and Neilson 2007). These guidelines assume that you have already reviewed and put into place the "first steps for any kind of aggression" listed in the previous section.

1. Identify the trigger (i.e., what event, item, situation, etc., is causing your dog to behave aggressively). There may be only one trigger, or there may be more than one. Identifying the trigger is important, as it allows you to reduce risk (by avoiding exposing the dog to this trigger while you work on changing the behavior). In addition, understanding the cause or causes of your dog's behavior will help you and your behaviorist or trainer (Box 7.1) to develop a targeted treatment plan aimed at reducing the behavior.
2. When working with a behaviorist or trainer, be sure that you understand the steps involved in changing the behavior. Any reputable behaviorist or trainer should be willing and able to explain the reasons for, and how to do, each step. The plan should be something that you can work with, given your own honest self-evaluation of the amount of time you have in your schedule, your level of comfort and skill in training your dog, and the number of other risk factors in your household (such as small children, other dogs, etc.). No matter how technically sound the behavioral modification plan, if you are unable to follow it, you will need to revise the plan and/or get hands-on help from a skilled trainer.
3. As noted above in the section on safety measures, you should have a firm plan in place for how you will manage the dog's environment to ensure safety of all involved. This may include changes to the dog's access to certain places (using dog crates or baby gates), muzzles, having him drag a long leash whenever he is not on a regular short leash (as it is much easier, and safer, to grab a long leash to regain control of a dog than it is to grab the dog himself), and so on. Your veterinarian may recommend neutering your intact male dog if he is behaving aggressively, or may recommend a daily medication (usually an anti-anxiety medication, or antidepressant) designed to help calm your dog while you work on changing his outlook and behavior.
4. Remember, for the reasons noted above, to use only positive reinforcement and rewards-based training to work with your dog. This lowers the risk to you and others interacting with your dog, improves efficiency of training, and reduces risk of increasing fear in your dog (as fear often contributes to aggressive behavior in dogs). For most dogs, using food in training is a proven and highly effective way to reinforce appropriate, desired behaviors (and, you can reduce daily meal amounts to accommodate the additional calories your dog consumes during training).
5. Although we have talked extensively about safety measures and environmental management for aggressive dogs, the main component of the approach is behavioral modification work to *reduce* aggressive behaviors. This approach is science-based; it uses established learning procedures in dogs, to change not just the dog's behavior, but also to change her underlying emotions and motivations in the presence of the trigger. Instead of a dog who reacts fearfully and/or aggressively in a given situation, we are working towards a dog who reacts positively, showing either happy anticipation or calm acceptance in the presence of the trigger.

Box 7.5. Foundation behaviors

These are behaviors that should be solidly in place with your dog before you begin to do your systematic desensitization/counterconditioning work to change his emotional response to his triggers, and/or as you work on response substitution to train and reinforce alternate, more desirable behaviors in the presence of his triggers. There are many different behaviors that could be considered "foundation behaviors" for this work (and in fact, some trainers define "foundation behaviors" as canine basic life skills such as sit, stay, and come when called). In general, for aggression treatment protocols, foundation behaviors include those behaviors that encourage and reinforce calm, focused (i.e., on you, vs. on the trigger) behaviors in your dog. Remember to begin by training these behaviors in an area with few if any distractions, and away from any known triggers.

Train behaviors to help increase and reinforce calm: Sit, stay, (lie) down.

Train behaviors to help you keep your dog's attention focused on you: Watch me, touch, "the name game" (reinforcing eye contact with your dog; i.e., say your dog's name, and every time he looks at you in response, mark and reward this behavior).

Train "emergency" behaviors to help get your dog out of situations where he unexpectedly encounters a trigger, or is over threshold for any reason: "Let's go" (fast walk in an alternate direction), emergency U-turn, and even an emergency recall (this would use a different cue than your regular "come when called" cue, and should mean, from the dog's perspective, "if you come back to my side right now, you will get a really great reward, each and every time!").

Resources

Emma Parson's great book, "Click to Calm," has a chapter on "Laying the groundwork: Foundation behaviors" that gives easy-to-follow instructions for how to train many of these behaviors with your dog, as well as when to use them.

Dr. Sophia Yin's website has a page with illustrations and video showing how to teach some of the more active techniques for redirecting your dog when she is over threshold, and for quickly getting her out of risky encounters: https://drsophiayin.com/blog/entry/reactive-dog-foundation-exercises-for-your-leash-reactive-dog/

Begin work on changing your dog's behavior by teaching what are often called the "foundation behaviors" (see Box 7.5 for descriptions of skills and resources for additional information). Foundation behaviors range from basic "life skills" like "sit" and "watch me," to more advanced, such as the "emergency U-turn" (moving quickly away from an approaching trigger). These are skills your dog will need to succeed during later stages of the work, during which you will begin to gradually expose her to the trigger that causes her aggressive behavior. Begin work on these skills in the absence of known triggers for your dog's behavior (so, these are good behaviors to work on during the early "avoid the triggers" stage of your work). A calm, focused dog will be much more able to learn new behaviors than a dog who is already reacting to the presence of something that upsets her.

6. In addition to teaching your dog these foundation behaviors, it may in some cases be helpful to make some changes in your daily interactions with your dog. These are particularly helpful if your dog is highly energetic or rambunctious, or if you have been perhaps a bit remiss

in training your dog in basic self-control and life skills – it happens more often than we'd like in our busy lives! Or, perhaps you have recently adopted or rescued a dog who has not lived in a home before, or whose previous owner was remiss or unsuccessful in training, resulting in a whirlwind of canine energy and boisterousness in your home. The goal here is to make interactions with your dog as clear and predictable for your dog as possible, so that she understands what is expected of her in any given situation, and she knows just what type of behavior will earn her a reward (vs. behavior which will not work for her, will not earn her any reward). Remember to avoid confrontational approaches during these interactions, as these may actually make the problems worse; instead give clear, consistent cues for desired behaviors, and reward correct responses to these cues. One recommended approach for restructuring interactions with your dog is the "nothing in life is free" approach (discussed earlier in this chapter), an approach which I have found very helpful in improving canine self-control. Patricia McConnell's short booklet, "How to be the leader of the pack,"[6] provides a great framework and practical tips for humanely providing boundaries for your dog's behavior.

7. Once you have put safety measures into place and worked on getting the necessary foundation behaviors in place, you can begin to again expose your dog to her triggers, in a gradual, controlled manner. This is where the learning processes of *systematic desensitization and counterconditioning*, described earlier (see also Figures 7.1 and 7.2), come into play. The goal, again, is to gradually change your dog's underlying emotional reaction to the triggers, which in turn will change (i.e., reduce!) her motivation to behave in an aggressive manner. In addition, use *"response substitution"*: use rewards-based training to teach your dog an alternate, preferable behavior to do in response to the trigger. Box 7.4 provided a list of recommended resources for dealing with canine aggression, but as a case study of how to set up and undertake a plan for reducing aggressive behavior, you might start by looking at Drs. McConnell and London's booklet, *Feisty Fido*. This very easy-to-read booklet presents an overview and practical how-to tips on dealing with one of the most common forms of aggression in dogs (leash reactivity).

On a final note, although aggression can be very stressful, in the vast majority of cases approaches and techniques like those discussed above can help reduce the behavior and bring back the joy of life with your canine companion. If you are dealing with any kind of serious aggression in your dog, take steps now to prevent escalation of the behavior, and to prevent bad events from happening. This is important, for your own sake as well as your dog's. Aggression is a behavior, usually performed for (from the dog's perspective, at least) a purpose. It isn't a value judgement; it doesn't mean a dog (or an owner) is "bad." The behavior may, or may not, be due to something done by humans; there may be underlying neurological or physiological issues contributing to the behavior, despite the best intentions of all who interact with the dog. In the majority of cases, this behavior can be changed, using the techniques described above, and with the help of a qualified professional. The risks must be considered, of course, and in some cases where

---

6 McConnell, P. (2007) *How to be the Leader of the Pack ... and have your dog love you for it!* (3rd ed.). McConnell Publishing Ltd. 16 p.

risks are very high, or prognosis for changing the behavior is very poor, difficult decisions must be made. Given the future prospects and quality of life for a dangerously aggressive dog (not to mention the stress, physical and financial risks for the dog's caretakers), arranging for humane euthanasia (as heartbreaking as that may be) may be the most humane decision.

Fears, phobias and anxieties

Fear is defined as an aversive (unpleasant) emotional state, a physiological and psychological response to a real threat. In this sense, fear can provide an adaptive advantage – it can help alert us to the presence of danger, and help us get out of a dangerous situation. Anxiety, on the other hand, is a more diffuse, generalized feeling of apprehension, or the anticipation of dangers from unknown origins; these anticipated dangers may be real, or imagined. Anxiety results in the same physiological reactions associated with fear, but may occur in the absence of any real or immediate threat. A state of anxiety can result from experiencing a fear-producing event, or from unpredictable environmental changes (most of us are calmer when in an environment we understand and can predict). A phobia, on the other hand, is a persistent, often excessive fear of a very specific stimulus – a fear of spiders, for example, or airplane rides, or veterinarians. Phobias can result from a traumatic experience (so, can be a learned behavior), but some phobias appear to be instinctive, ingrained fears of stimuli or situations that are inherently dangerous (such as fear of heights).

So, fear in and of itself is not necessarily a bad thing. Most animals, dogs included, are initially wary or afraid when presented with a novel stimulus, a new thing or situation with which they are not familiar. This is particularly true if that stimulus is very loud or intense (Hart et al. 2006). After repeatedly experiencing this stimulus, however, animals will usually habituate, and the fearful responses will be reduced or eliminated. Puppies are "primed" to learn about new things in their environment, and will generally habituate faster to new stimuli than adult dogs (see Chapter 5 on training for more on socialization and puppy development). For example, the first time a dog hears a thunderstorm, she may be fearful of the noises and flashes of light; however, after experiencing a number of such storms (during which nothing bad happened to her), she will eventually show only mild reactions to storms (Hart et al. 2006).

The problems with fear and anxiety in dogs occur when individual dogs are not able to habituate to a given situation or stimulus, resulting in continued fearfulness and associated behaviors (destructiveness, barking/whining, escape attempts, or even defensive aggression). Fear-related problems are much more commonly seen in dogs who were not well socialized as puppies; the period of socialization is when most dogs learn which things in the world are safe, and which are not. If they are not introduced and socialized to a stimulus during this period in their development, they are more likely to view it as "scary" later in life. Other factors which can influence fearfulness in adults can include traumatic experiences early in life and early maternal care (Foyer et al. 2016). Even the mental state of the mother during pregnancy may be important in determining the behavior and temperament of her puppies. It is well established, in a number of animal studies, that the stress levels experienced by the mother during pregnancy can affect neurotransmitters in her offspring, particularly serotonin metabolism, and can negatively influence behavior later in life (Peters 1986). In short, "stress-related events that occur in the perinatal period can permanently change brain and behaviour of the developing individual" (Maccari and Morley-Fletcher 2007, p. S10).

So, with all these potential influences on the level and incidence of fearfulness in our dogs, it is easy to conceptualize why some dogs seem so able to deal calmly, even happily, with all that life throws at them, while others seem to develop fearful behaviors and pathologies with only the slightest of provocations (and, every possible state between these two extremes). When working with veterinary students on behavioral issues with their own dogs, I had a number of very dog-savvy students express confusion and frustration about their recently-adopted island dog's level of fearfulness, despite raising this dog in the same way that they had raised previous puppies (with the previous puppies growing up to be calm, confident adults). Some of these differences, of course, may have been due to the different (and much more stressful, given the difficulty of the veterinary curriculum) lifestyle these students were now living, or other differences in the dog's environments. However, much of the difference may also have been due to the physiology and development of the dogs themselves (see Grigg et al. 2016 for a discussion of possible behavioral differences in these dogs).

> *Although many individuals experiencing stressful events do not develop pathologies, stress seems to be a provoking factor in those individuals with particular vulnerability, determined by genetic factors or earlier experience.*
>
> (McEwen and Sapolsky, 1995, cited in Maccari and Morley-Fletcher 2007)

Regardless of the causes underlying the behavior, the goal of treating fear and anxiety in dogs with behavioral modification is to change the dog's underlying internal state, to replace an existing fearful emotional reaction to a stimulus with a more positive, anticipatory or calm reaction. When we are faced with a dog who displays fear that is having a significantly negative impact on his quality of life (or, on ours), then in keeping with the Fifth Freedom (see Chapter 2), we owe it to him to try and help him get past this fear. Exposure to chronic stress can be detrimental to health, longevity, and learning (Mendl 1999). The behavioral modification technique of *counterconditioning* (described earlier) is the way to accomplish this change in the dog's emotional state in the presence of the stimulus. In order to keep the dog under threshold (i.e., not fearful) during the work, counterconditioning is often done in combination with systematic desensitization.

It is important to remember at this point in the discussion that we may not always be able to understand *why* our dog is fearful of a given stimulus, but from a practical standpoint, this is irrelevant. Even if their fears do not seem reasonable to us, we can't simply tell them to not be afraid, any more than someone could simply cure our fear of spiders, or flying, or public speaking by simply telling us we have no reason to be afraid. Again, from a practical standpoint when treating these issues, the important thing is to recognize that our dog is experiencing this fear,

> *It is not the physical characteristics of an aversive stimulus, but rather the cognitive appraisal of that stimulus, which determines its aversive character and whether a state commonly described as stress is induced.*
>
> (Koolhaas et al. 1999)

and then putting together a plan to help change his underlying reaction to that stimulus. Unlike when teaching your dog basic life skills, during which you are asking him to perform a given behavior (such as "sit" or "stay"), in this process you are essentially trying to control his internal, emotional state in a given situation (McConnell 2005).

First steps: recognizing fear in your dog, and managing for safety

Just as, when dealing with aggression, it is important to be able to recognize when your dog is becoming tense or aroused, it is also important to recognize when your dog is afraid. Behavioral symptoms of fear in dogs can include hypervigilance (frequently looking or glancing around, as though looking for approaching threats), avoidance behaviors (attempting to look or move away from the stimulus), lip licking (also called "tongue flicking," Figure 7.3), dilated pupils, vocalization (whining, growling, barking), destruction (particularly when left alone[7]), or in some cases, aggression (i.e., defensive aggression seen when handled, approached, or restrained). Chapter 4 of this book covers canine body language in more detail, and is important reading before attempting behavioral modification with your dog.

In addition, as fearful dogs can sometimes be at risk of hurting themselves (or others, if they become defensively aggressive), it is important at this stage to put safety measures into place to protect your dog and any potential targets of defensive, fear-based aggression. Avoid exposing your dog to things that scare her, unless you are in the process of doing counterconditioning work (or, once you have succeeded in helping her overcome her fears). Avoid the use of positive punishments in these situations, as these will likely only increase (rather than decrease) her fearfulness. On the other hand, many experts caution against reassuring or praising your dog when she is behaving fearfully. Although from a technical standpoint, it may not be possible to reinforce fear with rewards (as the positive emotion associated

Figure 7.3: Lip licking in a nervous dog. (Photo: R. Hack/S. Anthony)

with the reward will displace, rather than augment, the fear), it may be possible to inadvertently reinforce the behaviors associated with fear. In other words, in some situations the dog

---

7 We will talk about a particular form of anxiety that occurs when the dog is left alone in the next section.

may learn that a good way to get your attention is to behave fearfully; for dogs who value our attention highly, this could be a strong motivator for continuing to exhibit fearful behaviors. Does this mean you should never comfort your dog when she appears fearful? Not necessarily; dogs do tend to look to their human companions for reassurance and security, and it can be hard to resist the natural urge to comfort them when they are worried. But, dogs also look to us to help them determine how to react to a situation (i.e., if mom is calm and relaxed, then there must be nothing to worry about). A better approach, therefore, may be to distract your dog in the moment with a fun activity or training exercise (if the stimulus is unavoidable); providing your dog with a safe refuge for him to retreat to (in the case of dogs who are fearful of visitors, storms, or similar); or simply moving away from the stimulus, in a calm and relaxed manner (and then beginning to devise and enact your plan, using counterconditioning and systematic desensitization, to reduce his fear).

Next steps: reducing fear in your dog

The most effective techniques for reducing fear in dogs are *systematic desensitization* and *counterconditioning*. Without revisiting the theory behind these approaches (as those are covered earlier in the chapter), steps to follow would include:

1. Train and reinforce calm behaviors, using operant conditioning, and beginning your training in the absence of the trigger. This could be as simple as calmly teaching your dog "sit," "stay," and "down," using rewards for successful responses to the cues, and for calm behavior during the session. Some behaviorists suggest establishing a "safe haven" that your dog associates with periods of calm, massage, or similar (Emily Levine, in ACVB's *Decoding Your Dog*, 2014). This "safe haven" could be a mat or bed that you then use during your systematic desensitization and counterconditioning work, to help your dog remain calm in the presence of the stimulus. The goal is for your dog to be able to have a pleasant experience in the presence of the stimulus, and through repetition of this occurrence, to learn that good things can happen in the presence of the (previously scary) stimulus.
2. Whatever the stimulus is that causes fear in your dog, establish a gradient for levels of this stimulus, and start exposures at the lowest possible level. Your dog should be calm during these exercises; if your dog becomes agitated at any time, then the level of the trigger is too high. In that case, you'll need to backtrack to a lower level of the trigger, at which your dog is able to regain calm, and work at that level until your dog is ready to progress to the next level. How do you know when he is ready to progress? Whatever the level you are currently working at, your dog should be able to consistently remain calm as you work on simple cues like "sit," "watch me," and "stay," showing only occasional or mild interest in the trigger. Once he is at this point, you can try increasing the level of the trigger. Remember to reward calm behaviors in the presence of the trigger, even if the dog offers them spontaneously (in addition to in response to a cue from you).
3. Identify a treat or reward that your dog values *highly* (this might be a favorite human food item, or for some dogs, a game of tug with a favorite toy, for example). Some popular trainer favorites

for high-value food rewards are shown in Box 7.6. Then, use this high-value reward (for training, and rewarding calm behavior) *only* in the presence of the stimulus. When the stimulus is not present, the high-value reward goes away, too. This is the counterconditioning part of the process, and is done to try and build a positive association between the stimulus and the high-value reward in your dog's mind. As noted earlier, the stimulus should eventually become a predictor of the reward (rather than a fear-inducing stimulus).

> **Box 7.6. Options for high-value food rewards for use in counterconditioning work**
>
> Food is an excellent motivator for the majority of dogs, and here are some tried-and-true trainer favorites for use in behavioral modification work. Check with your veterinarian if your dog has any food allergies, and remember you may need to decrease the size of your dog's regular meals if you are using food rewards regularly. Keep size of food treats small (so, for example, cut larger treats into pea-sized chunks), so that the dog can consume them quickly, and to avoid the dog becoming satiated too quickly.
>
> - hot dogs (turkey, chicken, beef, or pork)
> - stinky sausage, such as braunschweiger or liverwurst
> - hard, strong-tasting cheeses such as parmesan
> - string cheese sticks
> - chunks of cooked (boneless) chicken (better for dogs with sensitive stomachs)

4. Keep training sessions short (5 to 15 minutes max), but frequent – every day is ideal.
5. Gradually increase the intensity of the trigger, and take care to go at your dog's pace. Depending on your dog's temperament, how strong his emotional reaction is, and how long he has been experiencing this fear, this process may go quickly, but may progress slowly. Patience is the key to success, for many dogs. Although I've seen the process occur quickly, in a matter of weeks, for some dogs, it may take a year or so to significantly change the behavior.
6. Another useful technique for reducing fear and reactivity towards a specific stimulus that is popular with many trainers is the "look at that" game. In this technique, the dog is rewarded just for looking at (without reacting to) the stimulus, starting at a "safe" distance from the stimulus and gradually progressing to closer distances. For example, if you are using clicker training with your dog, you can initially capture this behavior by clicking the moment your dog looks towards the stimulus, and immediately rewarding her. If you are far enough from the stimulus that she is below threshold, she will in all likelihood look back towards you the moment she hears the "click," expecting her reward. Eventually, the behavior will progress to the point where your dog looks at the stimulus and then automatically looks back to you for a reward; essentially demonstrating that she has learned that the presence of the stimulus predicts a reward from you.
7. In cases where anxiety is having a significant detrimental effect on quality of life, your veterinarian may recommend an anti-anxiety or antidepressant medication. This may initially seem like an extreme measure, but these medications, when prescribed and monitored by a veterinarian, can be extremely helpful in restoring calm and good quality of life for your dog. Box 7.7 lists some of the more common medications used to treat fear- and anxiety-based behavioral problems in dogs, with tips on when these might be necessary to help your dog.

Box 7.7. Medications for reducing fear and anxiety in dogs, and tips for when they may be needed

The trainer and author Suzanne Clothier has a great article* on her website listing her tips for determining when a dog might benefit from therapeutic medication, and for when she would refer a client to their veterinarian to discuss whether medication is warranted for their dog. *Only a licensed veterinarian should prescribe these medications*, and they can help you understand how these medications work, any possible side effects, and when you should expect to see a beneficial effect. Clothier's assessment is based on what she calls the "three Ps": Provocation, Proportion, and Persistence:

Provocation: What is provoking the dog's reaction? Is the dog reacting to something that is in fact an ordinary noise, or routine occurrence? Loud and/or intense stimuli are likely to provoke a reaction from any dog, and any dog who is already over threshold may react to a relatively mild added stimulus. However, if the dog is reacting intensely to a mild, very common occurrence, this may indicate that medication to reduce anxiety will be helpful.

Proportion: Is the dog's response in proportion to the provoking stimulus? Any dog may react briefly to the sound of a book being dropped, or the garden gate banging shut, but if the dog responds to these sounds by cowering and panting in the corner of the room, this dog may benefit from an anxiolytic medication.

Persistence: How long does the dog's response persist? If the dog in the previous example cowers briefly at the sound of a dropped book, but then almost immediately bounces back, this is very different from a dog who cowers in the corner for an hour following the sound. When it takes a long time for the dog to recover from the stimulus, this may indicate that the dog would benefit from medication.

Medications commonly prescribed for treating behavioral problems in dogs

Benzodiazepines: Fast-acting, anxiety-reducing medications, often used on an as-needed basis, such as diazepam, alprazolam, and clorazepate.

Selective serotonin reuptake inhibitors (SSRIs) and *tricyclic antidepressants* (TCAs): These antidepressant and anxiolytic medications (sometimes called maintenance medications for their long-term use) are given daily, and take longer (usually four to six weeks) to take effect. Examples of these include fluoxetine, clomipramine, and amitriptyline.

One to avoid? Acepromazine was heavily used in the past to attempt to control fearful behavior and prevent defensive aggression in dogs. Use of this medication is no longer recommended by behaviorists and DVMs experienced in behavioral medicine, as while it can act as a sedative, it is not an anxiolytic, and the dog may remain highly anxious while under the effects of this drug.

*Clothier, S. Does Your Dog Need Medication? Available at www.suzanneclothier.com (accessed April 2016)

If the anxiety is at a lower level, you may find it helpful to try one of the better-known commercially available products designed to help reduce fear and anxiety in dogs. Box 7.8 lists some of these. Note that these products do not have the bulk of clinical research behind them that the medications do. However, some owners report marked improvement with use of these tools, so they are worth a try in cases of less severe anxiety, or (in some cases) as an

> **Box 7.8. Commercially available products which may be helpful in reducing fear or anxiety in your dog**
>
> The *ThunderShirt* (http://www.thundershirt.com/) works like swaddling a baby, providing firm pressure that appears to comfort some dogs.
>
> *Adaptil* (CEVA Pharmaceuticals; http://www.adaptil.com/us/) also known as dog appeasing pheromone, is a synthetic analog of a pheromone produced by a mother dog, that has a calming effect on her puppies. It has been shown to have a calming effect on many dogs, of all ages, and is available in body-heat-activated collar or plug-in dispenser formats.
>
> The *ThunderCap* (formerly, Calming Cap; http://www.thundershirt.com/thundercap.html) filters the dog's vision (although they can still see through it) to reduce visual stimulation. Many dog owners have reported this as a useful tool in reducing reactivity to visual stimuli to things like the sight of another dog, cats, and so on.

> **Box 7.9. Resources for treating fears and phobias in your dog**
>
> **Books**
>
> Rugaas, T. *On Talking Terms with Dogs: Calming signals*. On understanding the body language of stress in your dog.
>
> McConnell, P. *The Cautious Canine*. A must-read booklet on classical counterconditioning, and working with fearful dogs.
>
> Levine, E.D. "I know it's going to rain, and I hate the Fourth of July." Chapter 12 in ACVB (2014), *Decoding Your Dog*. Specifically focused on noise phobias (thunderstorm anxiety, etc.), with helpful tips on identifying noise phobias in your dog, and treatment tips.
>
> Dodman, N.H. "The fearful dog." Chapter 6 in Dr. Dodman's 2008 book, *The Well-Adjusted Dog*. A nice review of the reasons why some dogs become fearful, and an introduction to treatment approaches for reducing fear.

add-on or adjunct therapy to the more intensive approaches. In all cases, however, behavioral modification remains the cornerstone of treatments for treating behavioral problems in dogs; don't assume that the medications or commercial products alone will be an effective substitute for doing the work of behavioral modification with your dog.

Finally, Box 7.9 provides a list of some recommended resources for reducing fearful behaviors in your dog. As a good starting point, Dr. McConnell's short and easy-to-read booklet, *The Cautious Canine*, contains a great overview of how to use counterconditioning to treat fear issues in your dog.

But, what if I'm not sure if it's really fear, or just plain old aggression?

As fear in dogs often progresses to aggression (as discussed earlier), and as fear aggression as a defense strategy (once learned and practiced repeatedly) can start looking an awful lot like confident, assertive aggression, this is often not an easy question to answer. We cannot read our dogs' minds, to tell whether fear is a component in the aggressive behavior or not. However, based on my experiences and discussions with many of my experienced professional colleagues (not to mention a considerable amount of reading on the topic), I will confidently say that in most cases, fear is a component of canine aggression. Sometimes, by considering the context of the aggression combined with the dog's body language, we can hazard a guess as to whether the dog is confident or fearful *at that moment*. Regardless for the precise motivation (or motivations) for the aggression – which we may never really know – the best course of action is to promptly take steps to change the behavior. Begin the behavioral modification steps outlined above: safety protocols, teaching foundation behaviors (and adding structure to the dog's environment), systematic desensitization and counterconditioning, response substitution. These techniques are tried-and-true ways to improve both types of behavioral issue in dogs. Of course, dogs are not "black boxes," for which any approach must be followed rigorously and on a set timeline, regardless of the feedback that we get from our dogs. Every dog is an individual, and no one knows your dog better than you. Be patient and consistent, work through these steps at your dog's pace (paying attention to how your dog is doing throughout), and seek the help of a behaviorist and/or trainer when necessary. Use medication as necessary, if recommended by your veterinarian, to help your dog get over her fears or impulse control issues. Go with what works, and don't forget to reward your dog when she behaves well! Once you're on the right track, you'll make great strides towards improved behavior.

Separation anxiety

The final category of behavior problems we will touch on is separation anxiety. Separation anxiety is a specialized form of anxiety, in which the dog experiences marked distress when separated from their human companions (often the human, or attachment figure, to whom they are most closely bonded) (Overall 1997; Horwitz and Neilson 2007). Separation anxiety can range in severity from mild to severe, and can occur when the dog is truly alone, or when they perceive that they are in some way separated from their human companion. How do you know if your dog is experiencing separation anxiety? If you are regularly returning home to any of the following behaviors (Horwitz and Neilson 2007; Palestrini et al. 2010), your dog may have separation anxiety (we'll talk about how to *confirm* separation anxiety later in this section):

- destruction (often focused at exit points, such as doors and windows);
- excessive barking, whining (or, angry notes from neighbors letting you know this is occurring);
- self-inflicted trauma (dogs will sometimes injure themselves when panicked, often during escape attempts through doors or windows);
- urinating/defecating around the house (in an adult dog, or previously well-house-trained dog);
- escape behavior (signs that the dog has been attempting to escape from where he is confined, or actual escapes out into the neighborhood during your absence);
- panting/drooling/pacing (these are signs of high anxiety in dogs, and are primarily seen in videos of these dogs when left alone, which we'll talk about next; excessive drooling in your

absence is often evidenced by damp or soaked bedding, but with no accompanying smell, etc., of urine);
- anorexia (another sign of high anxiety in dogs; in this case, you could leave a bowl of your dog's favorite food on the floor for him upon your departure, and come home to the food untouched – although he may eat it immediately following your return).

Note that, if this is truly separation anxiety, you will only really see these behaviors when the dog is left alone (and, you will see them consistently when the dog is left alone). If your dog is happily destructive in your presence, chewing furniture and pillows or the like, or regularly making house-training mistakes while you are at home, for example, you are likely not dealing with true separation anxiety, but an alternate issue. If your dog has been perfectly happy for years when left alone, until one day you come home to signs of canine panic (destruction, inappropriate elimination), it is unlikely that your dog has suddenly developed separation anxiety. It may be worth investigating whether something could have happened that day during your absence that frightened your dog (loud noises are a common culprit). I once had a client in this situation, who later learned that her next-door neighbor had been jackhammering concrete that day as part of a garden improvement project. In dogs that do *not* have separation anxiety, there may be other explanations for the behaviors listed above. For example, a dog, and particularly a puppy, who is left home alone and uncontained for long hours may become bored and frustrated, and seek out satisfying objects to chew in order to relieve her stress and boredom in your absence. Similarly, a dog who eliminates in the house when you are not home may be physically unable to hold his urine, and so on, for the length of time he is being left alone, and may need a visit in the middle of each work day by a friend or dog-sitter, to take him out to eliminate; or, he may just need a remedial course in house-training (combined, perhaps, with crate training so that he can be contained when you are not at home). Your vet (or, most good puppy training books) can advise you on how long, based on age, health status, and so on, a dog is physically able to "hold it" before they will need to eliminate. That being said, however, it is possible for any dog to have more than one behavioral problem, so if you are consistently seeing these behaviors when your dog is alone, but seeing some or all of them (or, seeing other undesirable behaviors) when you are home, it may be worth investigating your dog's behavior when he is left alone.

How can I be sure that my dog really has separation anxiety?

But, how do we confirm that our dog has separation anxiety, as by definition, separation anxiety only sets in when we are not present? One common approach for confirming separation anxiety in your dog is to use video in some format, to record what your dog does while you are away (see Scaglia et al. 2013 and Palestrini et al. 2010 for examples of use of video for assessing separation anxiety). In the age of web cams, "nanny cams," digital camcorders and the like, it is not too difficult to find a way to set up a camera out of reach of the dog, but focused on the area where the dog spends his time (or, if he is not contained in a crate or similar, then perhaps on the room or area where you tend to find the most damage). Turn the camera on as you leave for the day, and then examine the footage on your return. This can be very informative in understanding your dog's behavior, as well as a useful tool to show your veterinarian or behaviorist to help them advise you on the best course of action. In studies of separation anxiety dogs left alone, the anxious behaviors begin within 10 minutes of separation (Palestrini et al. 2010), so you do not need to worry if your video recorder will not record for the entire duration of your absence.

What to look for in the videos of your dog? Anxiety, fear, and even panic may be immediately obvious in the body language of your dog as you watch these videos: whining, pacing, panting, drooling, or barking are common signs. The panic seen in some dogs is so extreme that owners find it difficult to even watch these videos. Increased, often repetitive, motion and activity is particularly characteristic of anxious dogs when left alone, and is in marked contrast to the inactivity exhibited by dogs not suffering from separation anxiety when left alone (Scaglia et al. 2013). Most of us with dogs who do not suffer from separation anxiety suspect that our dogs spend the bulk of their time, when we are not home, sleeping or lounging on the (perhaps off-limits) comfortable couch. These video studies confirm that this is most likely true! If your dog, on the other hand, begins to whine, bark, pace, or scratch at the door soon after your departure (and, continues this behavior for an extended period of time), then your dog may indeed suffer from separation anxiety.

What causes separation anxiety? In one sense, given the strong bond between domestic dogs and their humans, and given the social nature of dogs, it makes intuitive sense that a dog separated from their canine or human companions would become anxious. If you are an animal used to living in social groups (whether with other dogs, or in human households), being alone in the world may not be the safest or most comfortable place to be. But, why then do some dogs suffer from separation anxiety, when others appear perfectly calm and happy when left alone? Why are some dogs able to learn to relax and rest when alone, while others appear unable to do so, to the point where they will sometimes injure themselves in apparent attempts to reunite with their human companion? Science historian Laurel Braitman, in her wonderful book, *Animal Madness*, writes movingly about her dog Oliver, who suffered from (among other things) severe separation anxiety, of watching video of Oliver in a full-blown panic at being left alone, and of the day when Oliver threw himself out of the window of their fourth-floor apartment, nearly killing himself in the process (Braitman 2014). For some dogs, separation anxiety truly is a life-threatening condition.

At this point, we don't really know what causes some dogs to exhibit separation anxiety while others don't. To some degree, genetic and physiological characteristics (many of which are beyond our control) likely contribute to the occurrence of the behavior, as with other anxiety-based behavior problems. Earlier descriptions of the behavior listed "risk factors" for the development of separation anxiety, such as time spent as a stray, in a shelter, rescue group, and so on; having a single owner; dogs who are "clingy" or display excessive greeting behaviors upon the owner's return. In many cases, later studies did

> Dogs who were adopted as adults, strays or rescues were not more likely to have separation anxiety when compared with the numbers of adults, rescues and strays in the overall hospital population. There appeared to be no one 'right' breed or source of dogs that would guarantee dogs free of separation anxiety.
>
> (Christensen and Overall, in *Decoding Your Dog*)

not support these generalizations (ACVB 2014). Many factors are difficult to really tease out, in a "which came first, the chicken or the egg" fashion. For example, a dog adopted from a shelter has separation anxiety; did she develop separation anxiety due to the stresses of living in the shelter? Or, was she relinquished to the shelter because she had separation anxiety? Or maybe, the traumatic separation from her previous family triggered the behavior, not the shelter itself? One recent study found that dogs separated from their litters at earlier ages (30 to 40 days) had a higher incidence of behavior problems such as destructiveness, excessive barking, attention seeking, fearfulness, and reactivity, than dogs who remained with their litters until 2 months (Pierantoni et al. 2011). In other words, leaving puppies with their mothers and littermates for eight weeks, during which time they can learn about the world from the secure base of their mother's side (and in contrast to the four to six weeks allowed by many US breeders), results in a more stable dog with a lower likelihood of developing separation anxiety. Nonetheless, this is likely only part of the picture – we are still learning what factors contribute to separation anxiety.

Is this behavior my fault?

In the past, owners of dogs with separation anxiety were often told that their own behavior had caused this problem. Owners were told that allowing the dog to sleep on the bed with them caused the problem, or that they didn't pay enough attention to the dog (or conversely, that they were paying too much attention and "spoiling" the dog). In some cases, owners were told that their dog was "angry" at them for leaving them alone, and acting out of spite or revenge (this is completely unsupported by the science). Owners were told to crate their dogs (which, while it may or may not reduce the damage caused by the dog, may actually make the anxiety worse and may result in injury as the dog attempts to escape the crate), or to get another dog (this last recommendation will be helpful for some dogs, but definitely not all dogs). Separation anxiety likely has a number of causes, which will vary dog to dog, just as it has a range of severity, and differences in how the anxiety manifests itself. Some dogs, for example, will be fine as long as there is any human, or even another dog, in the same house or enclosure with them. Other dogs will experience marked anxiety if their favorite person is not present, regardless of who else is. Owners may have had little or nothing to do (either deliberately or inadvertently) with the development of separation anxiety in their own dogs.

However, owners can certainly be a part of the solution. With patience, you can markedly improve and even eliminate this behavior problem, by gradually teaching your anxious dog that it is OK to be left alone. This is accomplished using some of the behavioral modification techniques we've already talked about, along with some specific strategies for addressing separation anxiety. For many dogs, particularly those with severe separation anxiety, anti-anxiety medications will be an important part of this process; see Box 7.7 for a list of some of the more commonly-used drugs, and talk to your veterinarian to see if pharmaceutical intervention would be helpful for your dog. For some dogs, the milder interventions listed in Box 7.8 may also be helpful.

A note on crates: in general, crates are a useful way to contain your dog during training, while traveling or at the vets, and so forth, and many dogs find their crates to be a safe refuge (and will voluntarily and happily spend time there). I always recommend that clients with new puppies get them used to being contained in a kennel or crate (using only systematic desensitization and rewards-based training methods, of course).[8] However, if your dog is fearful of being contained in a crate, or has to be dragged into the crate (never a good approach, even for dogs *not* suffering from separation anxiety), then being contained in a crate will not help solve separation anxiety, and will likely make the behavior worse by increasing anxiety. In addition, containment may only temporarily reduce collateral damage to property, and so on, as panicked dogs will go to great lengths to escape such confinement.

What to do: treating separation anxiety[9]

1. As with any behavioral modification plan, the first step is to devise ways to protect your dog from self-inflicted harm, and put these strategies into action immediately.
2. Management, to reduce anxiety and reduce opportunities for the dog to practice the behavior, should also be put into place. For example, early in the process of treating your dog's separation anxiety, avoid long separations to the greatest extent possible. In reality, most of us have to go to work each weekday, and may not be able to avoid such separations; in this case, try hiring a dog-sitter, enlisting a friend who works from home, or enrolling your dog in a "doggy day care" facility, to avoid leaving him alone for long periods.
3. Ensure that your dog has sufficient exercise, particularly before you leave the house (and by "sufficient" I mean a long walk or run, not a 5-minute foray into the back garden). Exercise (in dogs as in humans) can help reduce stress, and leave your dog more tired when you are not home. This alone will not solve the problem, but can be helpful.
4. Take the emotions out of greetings, and goodbyes. When leaving, behave as calmly as possible, as though nothing out of the ordinary was occurring. Ignore your dog for 10 to 20 minutes before departure. Making a fuss of your dog just before you leave may actually exacerbate the problem, by getting your dog worked up before you leave, or by inadvertently signaling that you will be gone for a long time. Similarly, upon returning, greet your dog in a calm manner; if he is overly excited at your return, ignore him for a few minutes until he is calmer, and then say your hellos. In some cases, your dog may learn "pre-departure cues" – signs (such as picking up your keys, or putting on your coat) that you are getting ready to leave – and may begin to become anxious at these signs. If this happens, you may need to do some desensitization for these cues; for example, at times when you are not planning to leave in the next few hours, make a point of picking up your keys, rattling them briefly, and then simply putting them back down. Repeat this periodically, and until your dog no longer pays any attention to the sound of your keys. You can also try using counterconditioning here: pair the departure cue with a favorite food treat or toy by presenting it at the same time as the departure cue; when the cue stops, so do the treats. If using a long-lasting food treat like a stuffed Kong toy, try removing the Kong when you put your keys away, waiting a

---

8 Crate training is covered in Chapter 5.
9 Recommendations adapted from Horwitz and Neilson 2007, ACVB 2014, and McConnell 2000.

moment, and then giving it back to him as you jingle your keys again. Eventually, present the departure cue *just before* you present the food reward, so that the departure cue becomes a predictor of good things ("I just heard the keys! That must mean I'm about to get a Kong toy filled with peanut butter …"). The goal here is to reduce the build-up of anxiety associated with the departure cue, by replacing the negative association (i.e., with your leaving), with a positive association (i.e., with a favorite food treat or toy).

5. Do not punish your dog if you return home to discover damage. This is not helpful, as your dog may not understand what he has done wrong, and such punishment may in fact make your dog more anxious, not less. The primary goal in treating separation anxiety is to reduce anxiety in the dog. It goes without saying, then, that you should avoid doing things that are highly likely to increase fearfulness and anxiety. And, from an ethical standpoint, punishing a dog for being afraid is not kind; your dog may not be able to control his behavior when he is in full-blown panic at being left alone. Instead, clean up the mess, take steps to prevent it from happening again, and keep on working on your treatment plan.

6. Help your dog to develop her own independence, and reinforce calm behaviors, at all times. Teach your dog, using rewards-based training, to sit/stay, lie down, to go to her mat or bed on cue. Over time, work on building her ability to remain calm while separated (physically) from you. For example, you may start working on "sit on your mat" at your side, if this is where she is most comfortable, but over time, gradually shift her mat further across the room, then into the next room (although with the door open so that she can see you), gradually moving her mat further and further away. Eventually, you can begin briefly closing the door between her and where you are, gradually increasing the amount of time she is "alone" in another part of the home. This is, of course, systematic desensitization; as always with this approach, it is important to move at the dog's pace, only increasing the stimulus (in this case, distance and/or duration of separation) when your dog is ready, as evidenced by calm behavior at the level you are working at prior to the increase. Reward spontaneous calm and/or independent behaviors when you see them, and ignore attention-seeking behaviors (to avoid reinforcing fearful, anxious behaviors).

7. Whenever you must leave the house, present your dog with a high-value food treat, such as the Kong stuffed with peanut butter or cheese paste, to work on while you are away. Many dogs, particularly those with milder forms of separation anxiety, will be distracted by this for a significant proportion of the time that they are alone.

8. As with most fear- and anxiety-based behaviors, systematic desensitization is one of the most effective approaches to resolving this behavior problem. Start by practicing very brief departures (30 seconds to a minute, being sure that your dog is able to remain calm for this time period), returning after this time; as noted above, treat these departures as non-events by essentially ignoring your dog, do not attempt to reassure her during these exercises. Repeat this brief separation until your dog is relaxed and calm, even uninterested, throughout the exercise; then, increase the duration of separation slightly. As you increase the duration of these training separations, occasionally revert temporarily to a shorter separation, to help reassure the dog that you will not be gone for too long. Vary the time of day during which you conduct your training sessions, to help your dog to generalize what she learns to any time of day. Given the reality of our lives and the unavoidable separations that occur for

work, family commitments, and so on, many behaviorists recommend using a novel cue of some sort to signal to your dog during all training (i.e., brief) separations. For example, choose a sound or scent that the dog has not been exposed to before: a spritz of lavender oil air freshener, a ring tone on your phone that you never use, a classical music CD, or a particular (soothing) sound from a white noise machine or app. Once you have chosen this training departure cue, always use the *same* signal on all training departures. Eventually, once the dog is able to remain calm during training departures of sustained duration, you can begin using this for "real" departures as well.

9. Your veterinarian may recommend using an anxiolytic medication to help with this process, particularly if your dog's separation anxiety is severe. The goal of using these types of medications is to help your dog to remain calm in the face of stressors, and thus be able to learn that *nothing bad happens when she is left alone*. This goal will be very difficult to achieve if your dog is so anxious that even short separations from you cause her great distress. In these cases, something bad *is* happening to her during these separations: she is experiencing the aversive emotions of fear and panic. Talk to your veterinarian to see if an anxiolytic medication would be helpful for your dog in regaining her peace of mind.

Patience is required during this process, and it may be helpful to remind ourselves periodically that our dogs are not performing these behaviors out of spite or as revenge for being left alone, they are genuinely struggling with their own fears. Separation anxiety can be very frustrating, but in many cases, responds well to behavioral modification work, and use of medication can greatly increase the speed of improvement. Box 7.10 presents some useful resources for reading more about separation anxiety in dogs, and for additional treatment suggestions.

Box 7.10. Resources on treating separation anxiety in your dog

McConnell, P. (2000) *I'll be home soon! How to prevent and treat separation anxiety*. A very short, easy-to-read booklet, which presents a practical treatment plan.

Christensen, E., Overall, K.L. (2014) Loyalty gone overboard: Separation anxiety. Chapter 11 in ACVB's *Decoding Your Dog*. A more in-depth look at signs and possible causes of separation anxiety, and treatment tips.

Yin, S. (2010) *How to Behave So Your Dog Behaves*. Dr. Yin's book has a short chapter on separation anxiety, as well as other chapters on crate training, tips for reinforcing calm behaviors, and links to her "learn to earn" program which can provide structure for anxious dogs.

A final thought on behavioral issues and quality of life: realistic expectations are important

Mental wellness is as important as physical wellness in ensuring optimal quality of life for our dogs. So, in striving to give our dogs the best possible quality of life, we owe it to them to try and help them to overcome their fears, to maintain calm and self-control, and to thrive in a world full of human expectations about their behavior. But, if we take on the responsibility of bringing a dog into our lives and homes, we also owe it to them to love them despite their faults (we all have a few, after all), and to appreciate them as dogs (not small, hairy humans). It's important to go into the process of behavioral modification with realistic expectations about your dog's ability to change, and to acknowledge that your dog may be struggling with these issues, just as you are. While you may get your dog who barks and lunges at other dogs to the point where she calmly walks past other dogs, even perhaps greeting them briefly before moving on, she may never be a dog who is happy or even reliable at the crowded, off-leash dog park. And, that's OK, if that is who your dog is. If you are working with an experienced behaviorist, they will be able to help you set realistic goals for your dog. Expecting too much, too fast, can be very frustrating for both you and your dog.

Take-home messages

1. 1. When you are faced with a serious behavioral issue in your dog, consult a qualified, experienced professional for advice and hands-on help. Get help promptly – the vast majority of behavioral problems will not resolve on their own, and treatment is much easier (and quicker) with early intervention.
2. Two important first steps in treating behavioral problems are:
    a. Consult your veterinarian: ensure that there is not an underlying medical problem contributing to your dog's behavior; and
    b. Consider what risks your dog's behavior presents, to humans, other animals in your household, and to himself. When necessary, put sensible safety precautions into place immediately to reduce or eliminate these risks, until you have successfully modified the risky behavior.
3. Just like training basic cues like "sit" and "come," behavioral modification for treatment of behavior problems is based on the science of how dogs learn, and the techniques used are based on these learning processes. These techniques are focused on helping these dogs to change their emotional reactions and learn alternate behaviors in the presence of their "triggers," using humane and rewards-based methods.
4. Two very useful behavioral modification techniques to be familiar with in treating behavior problems are:
    a. systematic desensitization: setting your dog up to gradually habituate to something in his environment that currently scares or upsets him; and
    b. counterconditioning: a form of classical conditioning, in which the goal is to replace an undesirable response (such as fear or aggression) with a more desirable one (such as calm or positive anticipation).
5. Don't forget about operant conditioning and rewards-based training (see Chapter 5) – these techniques can be your greatest allies in teaching and reinforcing the behaviors that you *do* want to see in your dog. Rather than attempting to punish the dog for an undesirable behavior, try instead to train a different, preferable behavior, and then redirect your dog when necessary (and reward him for the more desirable behavior).

References and additional resources

Aloff, B. (2002). *Aggression in Dogs: Practical management, prevention & behaviour modification*. Wenatchee, WA: Dogwise Publishing. 425 p.

American Animal Hospital Association (AAHA) (2015). 2015 AAHA Canine and Feline Behavior Management Guidelines. M. Hammerle, C. Horst, et al. (Eds.) *Journal of the American Animal Hospital Association* 51: 205–221.

American Animal Hospital Association/The Association for Feline Practitioners (AAHA/AAFP) (2015). 2015 AAHA/AAFP Pain Management Guidelines for Dogs and Cats. M. Epstein, I. Rodin, et al. (Eds.) *Journal of the American Animal Hospital Association* 51: 67–84.

American College of Veterinary Behaviorists (ACVB) (2014). *Decoding Your Dog*. D. Horwitz, J. Ciribassi, & S. Dale (Eds.) Boston, MA: Houghton Mifflin Harcourt. 384 p.

Beaver, B. (1994). Owner complaints about canine behavior. *JAVMA* 204: 1953–1955.

Braitman, L. (2014). *Animal Madness: Inside their minds*. New York, NY: Simon & Schuster Paperbacks. 379 p.

Christensen, E., & Overall, K.L. (2014). Loyalty gone overboard: separation anxiety. Ch. 11 in D. Horwitz, J. Ciribassi, & S. Dale (Eds.), ACVB (2014) *Decoding Your Dog*, Boston, MA: Houghton Mifflin Harcourt. pp. 235–262.

Dodman, N. (2008). *The Well-Adjusted Dog*. Boston, MA: Mariner Books/Houghton Mifflin Harcourt. 264 p.

Donaldson, J. (2004). *Fight! A practical guide to the treatment of dog–dog aggression*. San Francisco, CA: San Francisco SPCA. 116 p.

Foyer, P., Wilsson, E., & Jensen, P. (2016). Levels of maternal care in dogs affect adult offspring temperament. *Scientific Reports* 6, Article number: 19253.

Grigg, E.K, Nibblett, B.M., Sacks, B.N., Hack, R., Serpell, J.A., & Hart, L.A. (2016). Genetic and behavioral characteristics of the St. Kitts 'island dog'. *Applied Animal Behaviour Science* 2016. http://dx.doi.org/10.1016/j.applanim.2016.02.002.

Hart, B.L., Hart, L.A., & Bain, M.J. (2006). *Canine and Feline Behavior Therapy*. Oxford: Blackwell Publishing. 373 p.

Herron, M., Shofer, F.S., & Reisner, I.R. (2009). Survey of the use and outcome of confrontational and non-confrontational training methods in client-owned dogs showing undesired behaviors. *Applied Animal Behaviour Science* 117: 47–54.

Hetts, S. (2014). *12 Terrible Dog Training Mistakes Owners Make That Ruin Their Dog's Behavior ... And how to avoid them*. Animal Behavior Associates, Inc. 208 p.

Horwitz, D.F., & Neilson, J.C. (2007). *Canine and Feline Behavior* (Blackwell's Five-Minute Veterinary Consult Series). Oxford: Blackwell Publishing. 595 p.

Koolhaas, J., Korte, S.M, de Boer, S.F., et al. (1999). Coping styles in animals: Current status in behavior and stress-physiology. *Neuroscience & Biobehavioral Reviews* 23: 925–935.

Landsberg, G., Hunthausen, W., & Ackerman, L. (2013). *Behavior Problems of the Dog & Cat* (3rd ed.). London: Elsevier, Inc. 454 p.

Maccari, S., & Morley-Fletcher, S. (2007). Effects of prenatal restraint stress on the hypothalamus-pituitary-adrenal axis and related behavioural and neurobiological alterations. *Psychoneuroendocrinology* 32(Suppl.1): S10–S15.

McConnell, P. (2000). *I'll be home soon! How to prevent and treat separation anxiety*. Black Earth, WI: McConnell Publishing Ltd. 37 p.

McConnell, P. (2004). *Reading between the lines*. Black Earth, WI: McConnell Publishing Ltd./Tawzer Dog LLC.

McConnell, P. (2005). *The Cautious Canine* (2nd edition). Black Earth, WI: McConnell Publishing Ltd. 30 p.

McConnell, P. (2007). *How to be the Leader of the Pack ... And have your dog love you for it!* (3rd ed.). Black Earth, WI: McConnell Publishing Ltd. 16 p.

McConnell, P., & London, K. (2009). *Feisty Fido: Help for the leash-reactive dog*. (2nd edition). Black Earth, WI: McConnell Publishing Ltd. 63 p.

McConnell, P. (2010). *Treating Dog–Dog Reactivity*. Black Earth, WI: McConnell Publishing Ltd./Tawzer Dog LLC.

McEwen, B.S., & Sapolsky, R.M. (1995). Stress and cognitive function. *Current Opinions in Neurobiology* 5: 205–216.

McKeown, D., & Luescher, A. (1988). A case for companion animal behavior. *Canadian Veterinary Journal* 29: 74–75.

Mendl, M. (1999). Performing under pressure: Stress and cognitive function. *Applied Animal Behaviour Science* 65: 221–244.

Neilson, J.C., Eckstein, R.A.,& Hart, B.L. (1997). Effects of castration on problem behaviors in male dogs with reference to age and duration of behavior. *Journal of the American Veterinary Medical Association (JAVMA)* 211(2): 180–182.

Overall, K. (1997). Fears, anxieties and stereotypies. In: *Clinical Behavioral Medicine for Small Animals*. St. Louis, MO: Mosby-Year Book Inc. pp. 209–250.

Palestrini, C., Minero, M., Cannas, S., Rossi, E., & Frank, D. (2010). Video analysis of dogs with separation-related behaviors. *Applied Animal Behaviour Science* 124: 61–67.

Parsons, E. (2005). *Click to Calm: Healing the aggressive dog*. Waltham, MA: Karen Pryor Clickertraining. 181 p.

Peters, D.A. (1986). Prenatal stress: Effect on development of rat brain serotonergic neurons. *Pharmacology and Biochemical Behavior* 24: 1377–1382.

Pierantoni, L., Albertini, M., & Pirrone, F. (2011). Prevalence of owner-reported behaviours in dogs separated from the litter at two different ages. *Veterinary Record* 169(18): 468.

Pryor, K. (2002). *Don't Shoot the Dog! The new art of teaching and training*. Dorking, UK: Ringpress Books. 202 p.

Reisner, I.R. (2003). Differential diagnosis and management of human-directed aggression in dogs. *Veterinary Clinics of North America - Small Animal Practice* 33: 303–320.

Rugaas, T. (2006). *On Talking Terms with Dogs: Calming signals*. Wenatchee, WA: Dogwise Publishing. 79 p.

Salman, M., Hutchison, J., Ruch-Gallie, R., Kogan, L., New Jr., J.C., Kass, P.H., & Scarlett, J.M. (2000). Behavioral reasons for relinquishment of dogs and cats to 12 shelters. *Journal of Applied Animal Welfare Science* 3: 93–106.

Scaglia, E., Cannas, S., Minero, M., Frank, D., Bassi, A., & Palestrini, C. (2013). Video analysis of adult dogs when left home alone. *Journal of Veterinary Behavior: Clinical Applications and Research* 8: 412–417.

Stewart, G. (2012). *Behavior Adjustment Training: BAT for fear, frustration, and aggression in dogs*. Wenatchee, WA: Dogwise Publishing. 212 p.

Yin, S. (2010). *How to Behave So Your Dog Behaves* (2nd ed.) Neptune City, NJ: TFH Publications, Inc. 271 p.

Yin, S. (2014). *Dog Aggression: From fearful, reactive and hyperactive to focused, happy and calm*. Davis, CA: CattleDog Publishing.

Chapter 8

Canine physical wellness

Photo: iStock

A s a Great Dane enthusiast, I have always read that these gentle giants require much less exercise than you would expect from such a large dog, that in fact they are "couch potatoes" and perfectly happy as apartment dwellers. Someone forgot to explain this to my Great Dane, Aloysius (aka "Wishus"), who begged to race alongside of our mountain bikes up and down the amber hills of wheat country. He would charge up and over miles of trails with unbridled enthusiasm, until he was exhausted. As he aged, and we moved to a warmer climate, Wishus could not withstand the same level of exercise, but he always stayed active. Perhaps it was this athletic start that allowed him to live to the ripe old age of 12 (a pretty impressive age for a giant breed dog). Wishus had also started life as a victim of neglect and malnourishment prior to joining our family, but luckily this did not appear to inhibit his health (his only health issue was arthritis in his advanced age) or vitality. Perhaps it was this combination of both exercise and early calorie restriction that actually afforded him a long life. In this chapter we will explore how exercise and diet influence physical wellness as well as contribute to overall happiness in dogs. We will discuss exercise requirements and benefits, dog activities and sports you can enjoy with your dog, nutritional recommendations, types of diets, obesity, and how you can best create a life of physical wellness for your dog.

Exercise needs

It is clear that dogs require exercise, but just how much is beneficial and how much is required? Exercise is good for maintaining dogs' general health; it helps keep heart (Sessa et al. 1994), muscles, and joints healthy (Newton et al. 1997). Exercise can also improve behavior. According to Dr. Nicholas Dodman, in his book, *The Well-Adjusted Dog*, a tired dog is a good dog. Exercise is important for all dogs and it can be especially important in managing dogs with behavior problems. Dodman notes that different types of dogs require different levels of exercise. For instance, the sporting breeds require a lot of exercise, whereas toy dogs require less. The belief is that the amount of exercise required by a breed is dictated by the purpose for which that dog was originally bred. Dogs with a genetic history of covering long distances at a fast pace (such as huskies, malamutes, pointers, setters, spaniels, cattle dogs, and collies) need more exercise. Those bred for more sedentary lives (Bernese mountain dog, Chihuahua, Shih Tzu, French bulldog) require less vigorous exercise. This philosophy is held by a number of authors. If you search exercise needs for dogs, everything you will find will tell you that specific breeds require different levels of exercise. This is what I was taught as a student, and this is what I have often explained to clients. However, is this really the case? Dodman also emphasizes that each dog is an individual, and will have individualized needs for exercise. I do not think it is a coincidence that giant breeds like Great Danes are generally short lived, given the ideals we have for them such as massive size and low activity. These beliefs have disadvantaged these dogs in two ways, over-nutrition and under-exercise. There is evidence of a positive association between the perceived exercise requirements of dogs and dog walking. Dogs that are thought to require considerable exercise are walked more frequently, up to 150 minutes per week (Degeling et al. 2012). Given the proven benefits of exercise, perhaps we should rethink our ideas on exercise requirements in terms of the advantages to the individual dog, versus simply relying solely on breed recommendations. Of course there are physical limitations to the type and amount of exercise we should do with certain breeds, but instead of going by preconceived notions about historical breed exercise requirements, we should examine how each dog will benefit (and is benefitting) physically and mentally from different forms of exercise.

How much exercise does your dog need?

For the average dog, there is limited evidence on the ideal exercise requirements. According to the 2014 AAHA Weight Management Guidelines for Dogs and Cats (Brooks et al. 2014), with the exception of walking, there is little documented about the caloric expenditures of exercise.

You should consult your veterinarian before beginning a new exercise program for your dog, especially if your dog has any health conditions. The guidelines recommend: to safely start an exercise program, especially for the overweight and obese dog, begin with a 5-minute walk 2 to 3 times a day. You can gradually increase the length and intensity of the walk until either your dog reaches his limit (your dog should be happily tired, not exhausted, when you have finished exercising), or once a total of 30–45 minutes of walking per day has been achieved. Dogs will burn about 1.1 kcal/kg/km at a brisk walking pace of 10–10.5 min/km (Brooks et al. 2014). A 45 kg dog will burn about 240 calories after 4.82 km at that pace (Brooks et al. 2014). Walking at a slower pace also has health benefits, although the benefits are difficult to quantify due to lack of current data. Other types of exercise are undoubtedly beneficial as well, but there is a lack of research into the amount of other forms of exercise required to gain benefits equivalent or greater to regular walking. Monitor your dog and keep track of her progress. If you see any signs of fatigue or pain, always consult your veterinarian. Keep in mind that dogs with shorter legs, shorter snouts (brachiocephalic breeds like bulldogs and pugs), or other characteristics that may make them prone to overheating (such as Great Danes) may need extra monitoring for signs of exhaustion. For all dogs, mental stimulation is just as important as physical exercise. Try taking new routes with your dog and going to different dog friendly parks in your area. You can also introduce toys and play into your routine to keep walks stimulating for your dog.

Benefits of exercise

Exercise provides your dog with physical and psychological health. Exercise builds muscle, increases flexibility, burns fat, and strengthens the cardiopulmonary system. It may also stave off disease and aid in the treatment of conditions such as arthritis. An active lifestyle including exercise may even help prevent the onset of age-related cognitive decline or cognitive dysfunction syndrome (described in Chapter 10).

In humans, researchers have shown that the more a person exercises, the more brain cells they have in the frontal cortex of their brain (Cotman and Berchtold 2002). Additional benefits of exercise to human dog caregivers are well documented, including lower rates of coronary artery disease, blood pressure, stroke, type 2 diabetes, metabolic syndrome, colon cancer, breast cancer, and depression (Epping 2010). Evidence suggests that, in humans, as little as 1 hour per week of moderate to vigorous physical activity can reduce risks of a number of negative health consequences (Physical Activity Guidelines Report 2008, available at http://www.health.gov/paguidelines/). According to these guidelines, human adults need to engage in at least 150 minutes per week of moderate intensity aerobic activity (equivalent to a brisk walk for most people) or 75 minutes of vigorous intensity aerobic activity (an intensity at which it would be difficult to carry on a conversation).

Despite these substantial benefits and existing guidelines for activity, most Americans are not physically active at the recommended levels. Dog walking has been suggested as a way to help people get the recommended amount of exercise. In a recent study, dog caregivers in the US were monitored for the level and intensity of exercise they received while walking their dogs (Richards et al. 2014). The study revealed that dog walkers walked on average 1–2 times per

Figure 8.1: Exercise provides many benefits including increased physical health, mental health, and longevity. (Photo: © Donna Kelliher, www.DonnaKelliher.com)

day for an average of 30 minutes (plus or minus 15). Most of this time was spent in the moderate intensity activity range, which is consistent with the amounts of exercise recommended by the guidelines. A number of health and physical benefits are derived from regular brisk dog walking, and walking with a dog has been shown to help people adhere to an exercise regimen. A study in Australia found that dog owners walked more than non-dog owners (Baumann et al. 2001). Similarly, in a study completed at the University of Calgary, Canada, dog walkers were more than 10 times as likely to meet the recommended guidelines for human physical activity (Currie 2013).

In the United States, close to half the homes have one or more dogs (estimates range from 37–47 percent of homes; data from the American Pet Products Association, APPA, cited in ASPCA 2016). However, only half of those homes report walking their dogs. Despite health recommendations, very few dog caregivers walk their dogs for the recommended 30 minutes per day (Baumann et al. 2001). In a cross-sectional survey of 1,813 adults, Cutt and colleagues (2008) found that only 23 percent of dog caregivers walked their dogs 5 or more times per week.[1]

---

[1] However, compared to people without dogs, dog caregivers still completed significantly more minutes and sessions of walking and more minutes of total physical activity.

If dog walking is so beneficial, why are so few owners walking their dogs?

The strength of the bond between dog and caretaker is strongly correlated with how much dogs are walked. Feelings of attachment, companionship, and obligation within the dog–caregiver relationship are all correlated with high levels of dog walking. Community support and the availability of dog-supportive features in local parks, such as dog litter bags and bins, accessible water sources, fencing around designated off-leash areas, separation from children's play areas, dog agility equipment, and parks being well-fenced and not being located near busy roads, are all positively associated with dog walking (Westgarth et al. 2014). A primary barrier to dog walking is caregiver perception. Caregivers who do not walk their dogs report that their dog does not provide sufficient motivation to go on walks, and that they perceive social support provided to dog walkers as poor (Cutt et al. 2008). With the clear benefits of dog walking that have been outlined in numerous studies, caregivers should try to make exercise a priority; both caregiver and dog will benefit! In addition, communities should support these activities, in the ways noted above, for the welfare of their citizens (and their dogs).

Dog activities/sports

One way dog caregivers can feel more engaged in exercise for their dog is to participate in a dog activity or sport. This type of exercise can be fun for you, the caregiver, and your dog. In recent years a number of dog sports have been popularized for both the pet dog and the true competition dog. There are a variety of sports fit for every breed, personality, and energy level. Equally important, caregivers and their dogs can be competitive, or can simply participate for enjoyment and health benefits.

Agility

Dog agility is a dog sport in which a handler directs a dog through an obstacle course in a race for both time and accuracy. Dogs run off-leash with no food or toys as incentives, and the handler can touch neither dog nor obstacles.

Courses can contain tunnels, jumps, poles, and other obstacles. Courses are complicated enough that a dog could not complete them correctly without human direction. In competition, the handler must assess the course, decide on handling strategies, and direct the dog through the course, with precision and speed equally important.

Figure 8.2: Agility. (Photo: © Donna Kelliher, www.DonnaKelliher.com)

https://www.usdaa.com/se_agility.cfm – USDAA
http://www.akc.org/events/agility/ – American Kennel Club Agility
http://www.ukagility.com – United Kingdom Agility
http://agilityclub.org – The Agility Club

Disc dog

Disc dog is a sport in which dogs and their human flying disc throwers compete in events such as distance catching and somewhat choreographed freestyle catching. The sport celebrates the bond between handler and dog by allowing them to work together.

http://usddn.com – US Disc Dog Nationals
http://www.skyhoundz.com – Skyhoundz

Dock jumping

Dock jumping, also known as dock diving, is a dog sport in which dogs compete for distance or height when jumping from a dock into a body of water. There are dock jumping events in the United States and other countries such as the United Kingdom and Australia. Dock jumping first appeared in 1997 at the Incredible Dog Challenge, an event sponsored and produced by pet food manufacturer Purina. There are now a number of organizations that run dock jumping competitions in different countries.

http://northamericadivingdogs.com – North America Diving Dogs
http://www.ukcdogs.com/Web.nsf/WebPages/DogEvents/DockJumpingGettingStarted – United Kennel Club Dock Jumping

Field trials

A field trial is a competitive event at which hunting dogs compete against one another. There are field trials for retrievers, pointing dogs, and flushing dogs. Kennel clubs or other gun dog organizations usually organize field trials. Trials may vary dependent on requirements of different breed organizations.

www.ufta-online.com/ – United Field Trialers Association
www.ukbirddogs.com/field_trials.htm – UK Bird Dogs

Figure 8.3: Dock jumping. (Photo: © Donna Kelliher, www.DonnaKelliher.com)

Figure 8.4: Field trials. (Photo: © Donna Kelliher, www.DonnaKelliher.com)

Flyball

Flyball is a dog sport in which teams of dogs race against each other from a start line, over hurdles to a box that releases a tennis ball to be caught when the dog presses the spring-loaded pad, then back to their handlers while carrying the ball. Flyball is run in teams of four dogs, as a relay. Each dog must return its ball all the way across the start line before the next dog crosses.

>http://www.flyball.org – North American Flyball Association
>http://www.flyball.org.uk – British Flyball Association

Lure coursing

Lure coursing is a sport for dogs that involves chasing a mechanically operated lure. Competition is typically limited to dogs of purebred sight hound breeds, although there is an AKC pass/fail trail for all breeds called the coursing ability test.

>http://www.akc.org/events/lure-coursing/ – American Kennel Club Lure Coursing
>http://www.ukcdogs.com/Web.nsf/WebPages/DogEvents/LureCoursing – United Kennel Club Lure Coursing

Physical wellness

Figure 8.5: Flyball. (Photo: © Donna Kelliher, www.DonnaKelliher.com)

Figure 8.6: Lure coursing. (Photo: © Donna Kelliher, www.DonnaKelliher.com)

Musical canine freestyle/freestyle

Freestyle is a modern dog sport that is a mixture of obedience training, tricks, and dance that allows for creative interaction between dogs and their owners. The sport has developed into competitive forms in several countries around the world.

> http://www.canine-freestyle.org – Canine Freestyle Federation
> http://www.caninefreestylegb.com/index.htm – Canine Freestyle GB

Nosework

Nosework (also sometimes called "scent work") is a canine sport created to mimic professional detection dog tasks. One dog and one handler form a team. The dogs must find a hidden target odor, often ignoring distractors (such as food or toys), and alert the handler. After the dog finds the odor they are rewarded with food or a toy. Nosework is a fast growing sport in part because it accommodates canines with disabilities or behavior problems.

> https://www.nacsw.net – National Association of Canine Scent Work (NACSW)

Rally/rally obedience

Rally is a dog sport based on obedience. Unlike regular obedience, instead of waiting for the judge's orders, the competitors proceed around a course of designated stations with the dog in heel position. The course consists of 10 to 20 signs that instruct the team what to do. Unlike traditional obedience, handlers are allowed to encourage their dogs during the course.

> http://www.ukcdogs.com/Web.nsf/WebPages/DogEvents/RallyObedience – United Kennel Club Rally Obedience
> http://www.akc.org/events/rally/ – American Kennel Club Rally

Sheepdog trial/herding event/dog trial

Herding dog trials are a competitive dog sport in which herding dogs move sheep around a field, fences, gates, or enclosures as directed by their handlers. Such events are particularly associated with hill farming areas, where sheep range widely on largely unfenced land. These trials take place in the United Kingdom, Ireland, South Africa, Chile, Canada, the USA, Australia, New Zealand, and other farming nations.

> http://www.usbcha.com/ – United States Border Collie Handlers Association
> http://www.englishnationalsheepdogtrials.org.uk/ – International Sheep Dog Society

Sled dog racing/dog sled racing

Sled dog racing is a winter dog sport most popular in the Arctic regions of the United States, Canada, Russia, Greenland, and some European countries. It involves teams of sled dogs pulling a sled with the dog driver (or "musher") standing on the sled runners. The team completing the marked course in the shortest time is judged the winner.

> http://www.isdra.org – International sled dog racing association

Figure 8.7: Sheepdog trials. (Photo: © Donna Kelliher, www.DonnaKelliher.com)

This is not a comprehensive list of dog sports and activities, but a sampling of a range of activities in which dogs and caregivers can participate to get exercise and enjoyment, and to improve health and happiness. If there is an activity that you enjoy, there is probably a dog sport that will complement it.

Weight guidelines

An important aspect of canine physical well-being that is often overlooked is healthy weight. Determining and maintaining an ideal weight is one of the key factors in promoting a long happy life for your dog. Weight guidelines are variable and differ by breed. Weights listed by breed are often presented as a range, making it difficult to estimate just what is best for your canine companion. Additionally, these estimates may not be particularly accurate, depending on your dog's relative size and activity level. If your dog is a mixed breed, the estimates are even less accurate. Given these issues, the 9-point body condition score (often abbreviated BCS) was created as a tool for both veterinary professionals and caregivers to determine ideal body condition (Figure 8.8). The Pet Food Manufacturers Association (PFMA) provides a 4-point scale called the Pet Size-O-meter that offers similar, easy-to-use guidelines[2] (this body condition score has been

---

[2] The Pet Size-O-Meter and other obesity prevention information can be found at http://www.pfma.org.uk/_assets/docs/weigh-in-wednesday/pet-size-o-meter-dog.pdf

shown to be comparable with more invasive and comprehensive body composition measurement via DEXA, a dual x-ray scan used to monitor body mass (Jeusette et al. 2010; Bjornvad et al. 2011). The body condition score uses a two-step process to determine your dog's ideal condition: visual inspection and palpation (touch). After examining your dog, refer to the chart and assign your dog to one of three categories:

Under ideal (body condition score = 1–3)
Ideal (body condition score = 4–5)
Over ideal (body condition score = 6–9)

If your dog is not in the ideal condition range, work with your veterinarian to improve your dog's condition through diet and exercise. You can also determine the ideal body condition by determining muscle condition with muscle condition scoring (Figure 8.9) to give a more detailed view of your dog's condition.

You can determine a muscle condition score for your dog through visual examination and palpation (touching) over the spine, skull, scapulae, ribs, lumbar vertebrae, and pelvic bones. Muscle loss is typically first seen in the muscles on each side of the spine. You may see muscle loss at other sites but this can be more variable. The muscle condition is defined as normal, mild

Figure 8.8: Body condition scoring. Source: WSAVA Global Nutrition Committee Body Condition Score, available at http://www.wsava.org/nutrition-toolkit. (These guidelines were first published in the *Journal of Small Animal Practice*, July 2011; 52(7): 385–396, published by John Wiley and Sons Ltd and are published with permission. Global Nutrition Committee Toolkit provided courtesy of the World Small Animal Veterinary Association.) The WSAVA also has a helpful video demonstrating how to use the BCS, the *WSAVA Global Nutrition Committee – Body Condition Score Video*, at https://youtu.be/tf_-rwxqHYU

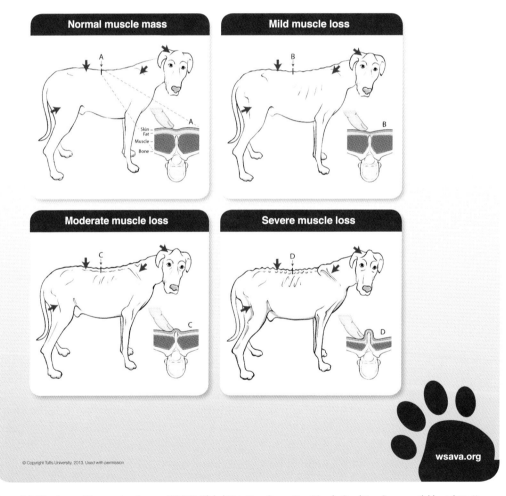

Figure 8.9: Muscle condition scoring. Source: WSAVA Global Nutrition Committee Muscle Condition Score, available at: http://www.wsava.org/nutrition-toolkit (These guidelines were first published in the *Journal of Small Animal Practice*, July 2011; 52(7): 385–396, published by John Wiley and Sons Ltd and are published with permission. Global Nutrition Committee Toolkit provided courtesy of the World Small Animal Veterinary Association.)

loss, moderate loss, or severe loss. Note that your dog may have considerable muscle loss even if he is overweight or can have normal muscle even when exceptionally lean. Therefore, assessing both body condition score and muscle condition score and assessing your dog as an individual is the best way to assure he is at his ideal condition.

Work with your veterinarian to monitor muscle mass and body condition to assure your dog is at its ideal weight and condition.

A dog's nutritional requirements

As our knowledge of the relationship between diet, exercise, and health continues to advance, and as the range of foods available for dogs continues to expand, it's more important than ever to understand the basics of dog nutrition and have an understanding of the nutrients that all dogs need to live a healthy, happy life. The following is a list of the dietary elements essential to all dogs.

Nutrients

Nutrients are substances obtained from food that a dog's body uses to function and that are necessary for maintenance and growth. There are six essential classes of nutrients dogs must have to remain healthy and happy (NRC 2006; Case 2014):

1. *Water*: Water is essential to a dog's life and accounts for the majority of a dog's body weight. Dogs must have access to water and opportunities for drinking at all times. Dehydration from lack of water can cause serious illness with as little as a 10 percent loss, or even death at a 15 percent loss of water.
2. *Protein*: Although dogs, unlike cats, are not true carnivores,[3] they cannot survive without protein. Proteins are the building blocks for cells, organs, enzymes, many hormones, and antibodies. Protein is obtained from animal products, some vegetables, and cereals. Proteins are made up of amino acids. Dogs cannot produce all the amino acids essential to life contained in proteins and must derive some of them from their diet. High-quality proteins have a good balance of all of these essential amino acids. The 10 essential amino acids for dogs are arginine, histidine, isoleucine, leucine, lysine, methionine, phenylalanine, threonine, tryptophan, and valine. Taurine is also considered essential for some dog breeds (spaniels, retrievers, and Newfoundlands) to avoid certain heart problems (Kittleson et al. 1997; Sanderson 2006).
3. *Fats*: Derived in the diet from animal fat and seeds, fats provide the most concentrated source of energy. They supply essential fatty acids that cannot be synthesized in the body

---

3   While it is true that dogs generally eat a high level of protein they also consume a significant amount of vegetable matter and grain through foraging and scavenging. In addition to eating grasses and grains, dogs consume the ingesta (undigested vegetable food matter) in prey. Therefore, dogs are best classified as (facultative) omnivores, since they cannot derive all the necessary nutritional requirements from meat. In comparison, cats are obligate carnivores, meaning that all of their nutritional requirements are met by a meat-based diet.

> **Box 8.1. Vitamins essential to the diet of a healthy dog**
>
> Vitamin K: activation of clotting factors, bone proteins, and other proteins
>
> Vitamin B1 (thiamin): energy and carbohydrate metabolism; activation of ion channels in neural tissue
>
> Riboflavin: enzyme functions
>
> Vitamin B6: glucose generation; red blood cell function; niacin synthesis; nervous system function; immune response; hormone regulation; gene activation
>
> Niacin: enzyme functions
>
> Pantothenic acid: energy metabolism
>
> Vitamin B12: enzyme functions
>
> Folic acid: amino acid and nucleotide metabolism; mitochondrial protein synthesis
>
> Choline: phospholipid cell membrane component
>
> *Sources:*
>
> The National Academies of Sciences, Engineering, and Medicine (2015). *Your Dog's Nutritional Needs: A science based guide for pet owners.* Washington DC: The National Academy of Sciences.
>
> Case, L.P. (2014) *Dog Food Logic.* Dogwise Publishing, Wenatchee, WA.

and they serve as transport for fat-soluble vitamins. Essential fatty acids are important to cell structure and are necessary for a healthy skin and coat. Deficiencies in the omega-3 family of essential fatty acids may be associated with vision problems and impaired learning ability. Fat also provides insulation and protection of internal organs, and cholesterol is the basic building block for steroid hormones.

4. *Carbohydrates*: Derived from plants and grains, carbohydrates typically make up anywhere from 30–70 percent of a dry dog food. They provide energy in the form of sugars and have fiber that is important to the health of the digestive tract. There is no minimum carbohydrate requirement but carbohydrates provide glucose that is needed to supply energy to organs including the brain. Fiber, both digestible and non-digestible, is a kind of carbohydrate that helps balance the bacterial population in the small and large intestines for the maintenance of good digestion.
5. *Vitamins*: Most vitamins cannot be synthesized in the body, and must be obtained in the diet (Box 8.1). Avoid dangerous toxicities by being careful to not over supplement your dog's diet with vitamins. If they are eating a high-quality commercially prepared diet, it will be complete and balanced with the appropriate vitamins. It is unnecessary to give a vitamin supplement unless your veterinarian diagnoses a vitamin deficiency. Over supplementation is more common than vitamin deficiency. For example, excess vitamin A may result in bone and joint pain, brittle bones, and dry skin. Excess vitamin D may result in very dense bones, soft tissue calcification, and kidney failure.

Box 8.2. Minerals essential to the diet of a healthy dog

Calcium: formation of bones and teeth; blood coagulation; nerve impulse transmission; muscle contraction; cell signaling

Phosphorus: skeletal structure; DNA and RNA structure; energy metabolism; locomotion; acid-base balance

Sodium: enzyme functions; muscle and nerve-cell membrane stability; hormone secretion and function; mineral structure of bones and teeth

Potassium: acid-base balance; nerve-impulse transmission; enzymatic reactions; transport functions

Sodium: acid-base balance; transfer of extracellular fluids across cell membranes, nerve impulse generation and transmission.

Chlorine: acid-base balance; transfer of extracellular fluids across cell membranes

Iron: synthesis of blood components; energy metabolism; regulation of osmotic pressure

Copper: connective tissue formation; iron metabolism; blood cell formation; melanin pigment formation; myelin formation; defense against oxidative damage

Zinc: enzyme reactions; cell replication; protein and carbohydrate metabolism; skin function; wound healing

Magnesium: enzyme functions; bone development; neurological function

Selenium: defense against oxidative damage; immune response

Iodine: thyroid hormone synthesis; cell differentiation; growth and development of puppies; regulation of metabolic rate

Sources:

The National Academies of Sciences, Engineering, and Medicine (2015). *Your Dog's Nutritional Needs: A science based guide for pet owners.* Washington DC: The National Academy of Sciences.

Case, L.P. (2014). *Dog Food Logic.* Wenatchee, WA: Dogwise Publishing.

6. *Minerals*: Minerals are nutrients that cannot be synthesized by animals and must be provided in the diet (Box 8.2). In general, minerals are most important as structural constituents of bones and teeth, for maintaining fluid balance, and for their involvement in many metabolic reactions.

Energy requirements

Energy is defined as the number of calories the food provides in a given volume or weight of food (Case 2014). There is no set amount or formula to calculate your dog's individual energy

Table 8.1 Examples of daily energy needs in calories (kilocalories*) per day.

	10 lb.	30 lb.	50 lb.	70 lb.	90 lb.
Puppies: 0–4 months	653				
Puppies: 4 months–adult	435	992	1455	1873	2261
Adult neutered/spayed dogs	348	794	1164	1498	1809
Adult intact dogs	392	893	1309	1685	2035
Working/active dogs	544	1240	1819	2341	2826
Obese/inactive dogs	283	645	946	1217	1470

*1 Calorie = 1 kilocalorie = 1,000. Energy needs are expressed in terms of kilocalories, which are equivalent to calories in this document. These values are examples of caloric requirements by life stage and activity level calculated from the Ohio State University Veterinary Medical Center Basic Calorie Calculator (http://vet.osu.edu/vmc/companion/our-services/nutrition-support-service/basic-calorie-calculator). These examples are only estimates, individual pet needs can vary by as much as 50 percent from calculated values. Please consult your veterinarian before changing your dog's diet.

needs, there are only estimates. There are numerous differences in requirements between breeds, size, life stage, activity levels, skin and coat thicknesses, body condition, and living conditions.

Table 8.1 provides some examples of caloric needs for dogs. This table is based upon a dog's resting energy requirements or RER. The RER is the basic amount of energy that a dog would use in a day while remaining at rest. The formula to calculate RER for dogs is:

$$\text{RER in kcal/day} = 70 \times (\text{BW [kg]})^{.75}$$

Any activity or variable (such as temperature change) other than rest will require an increase in energy (RER) and an increase in calories to meet the energy needs. Maintenance levels are not the same as resting energy levels. For example, for maintenance, a normal adult dog with average activity has a requirement for maintenance 1.6 times the resting energy requirement. Resting energy requirements vary based on life stage and activity level.

This table does not tell us how much we need to feed our dog but it does show the significant variability in a dog's energy requirements based upon his activity and life stage. This information can help us gauge when our dogs might require more food to maintain their body condition. There are a number of variables that are not taken into account in many tables like these, such as breed differences, ambient air temperature, and coat insulation, which can additionally alter an individual dog's energy requirement as much or more than any of the activities listed above. Please discuss calorie requirements and diet changes with your veterinarian.

Feeding for the life stage and energy level

Dogs are very efficient at processing their food, and rarely require energy levels above maintenance. The average dog or even one that competes in a dog sport generally can be maintained on a diet formulated for maintenance of an "average" dog. People often overestimate the number of calories burned during exercise (both human and animal), and so, overestimate the caloric needs of their dog.

Activity level

The canine athlete

These are the working dogs such as sled dogs or search and rescue dogs, or the highly competitive athletes such as those in top levels of agility or hunting trials that require daily training regimes. Some pet dogs may reach this level of conditioning if they are going on daily distance runs alongside their marathon- or distance-running caregivers. These are the dogs that regularly accompany their humans on long distance runs of over an hour and perhaps cover as many as 40 or more miles per week, or are regularly doing some other extreme activity. There are special diets formulated for these dogs that have increased levels of protein, carbohydrates, and fat, with nutrients that can be easily and quickly absorbed by their body. This type of diet is for the active times in the high performance dog's life. In the off-season, when these dogs are not maintaining these high levels of training, these dogs can be transitioned to a maintenance diet.

The weekend warrior

Dogs that spend most of their time during the week in the house or out in a yard but have bouts of increased activity sporadically or routinely on weekends can be fed at an adult maintenance level. These are the dogs that might participate in dog sports, go on long hikes or swims and the like. Although they may expend more energy than dogs walked occasionally, they tend to be efficient and require no more calories than the average dog. That said, if your dog cannot maintain a healthy body condition on a maintenance diet, you may need to increase his ration, or talk to your veterinarian about possible medical conditions contributing to his weight loss.

Life stage

There are only two life stages defined by the AAFCO (Association of American Feed Control Officials), the agency responsible for monitoring dog food label claims. One stage is "growth and reproduction" and the other is "adult dogs during maintenance" (Case 2014). This means that foods marketed for other life stages such as puppy, senior, less active, large breed, and so on, are not required to comply with an established or regulated set of standards for that life stage.

Puppies and large breed puppies

Puppies receive complete nutrition from their mother's milk for the first four weeks. After that time, puppies should be gradually weaned from their mother's milk and started on a high-quality puppy food. Puppies need about two to three times as many calories per pound of body weight as adult dogs of the same breed (Pet Food Manufacturers' Guidelines Breeders Newsletter, Summer 2010). Feed your puppy to body condition and be sure not to overfeed. Being overweight, even at puppyhood, can lead to health risks.

Large breed puppies are more prone to developmental skeletal disease when compared to smaller breeds because of their increased potential for rapid growth. The research shows that the primary reason puppies experience rapid growth is the overfeeding with nutrient-rich, energy dense foods during the rapid growth phase (normally between 3 and 6 months of age). Another contributor to developmental skeletal disease is the amount of calcium in the diet. Large breed puppies that consume too much calcium are at greater risk for developing skeletal issues such as hip and elbow dysplasia. The best solution is to feed a food that supplies the necessary nutrients with reduced energy (calories) and slightly reduced calcium. The diet of a large breed puppy food should contain a dry matter content of about 30 percent protein, 9 percent fat, 1.5 percent calcium, and 0.8 percent to 1 percent phosphorus. The calcium:phosphorus ratio should be between 1:1 to 1.3:1.

A common misconception is that dietary protein should be carefully regulated in large breed puppies to prevent skeletal abnormalities. In fact, this theory was disproved some years ago (Nap et al. 1991). Most commercial puppy foods contain more protein than is thought necessary, but studies have shown that protein contents of 23 percent to 31 percent (dry matter) do not have a deleterious effect on growth.

Another common mistake is to simply feed large breed puppies an adult dog food. While it may appear that this is a good solution for lower calorie density, adult dog foods vary wildly in their calorie content and they do not always contain the best calcium content or calcium:phosphorus ratio for puppies. If your puppy is a large breed, the best recommendation is to feed a large breed puppy food or a puppy food that contains all of the essential nutrients with slightly reduced energy and calcium.

In general, all puppies should also be fed a puppy food specifically formulated for their life stage. Raw puppy diets are more difficult to monitor for consistency and regulation of the calcium:phosphorus ratio. While they may contain healthful ingredients, there is no ideal way to determine the exact calories and calcium content of this type of varied diet. It can be estimated, but individual ingredients will vary. Bones are required in a puppy diet to assure that they are getting the correct calcium:phosphorus ratio (up to 10 percent of the diet) but it is difficult to determine how much bone can be safely added. The best way to assure good development is to not overfeed your puppy. Again, keeping your pup lean (but not too thin) is the best assurance of good bone development. There is a high correlation between plump pups and maximal growth rate, which is very risky for your dog. This is especially true for large breed puppies. Some caregivers of large breed pups express interest in having a very large/giant adult dog. Although we are tempted to have the adorable giant dog, think about how that maximal growth and size will influence their long-term health. Our emphasis on growing these giants is most likely a major contributor to their abbreviated life span.

Pregnancy and lactation

Pregnancy and lactation require a higher calorie diet. Some common mistakes that dog caregivers make when feeding a pregnant or nursing dog are to feed too much, too early in the pregnancy, and then to not feed enough during lactation. A dog's pregnancy lasts about 63 days and for the first four to five weeks they can be maintained on their regular maintenance diet. For the second half of pregnancy, however, dogs require 30–50 percent more calories to support their growing pups. You can talk to your veterinarian about transitioning your pregnant dog to a high-energy dog food or to a puppy formula. New mothers generally suckle their puppies for at least six weeks; the first milk given by mothers to their puppies (called colostrum) is essential to the puppies' development, particularly to their immune systems. The mother's need for calories increases with the number of puppies and the week of lactation, up to four weeks. Giant breeds (like Great Danes) have proportionately smaller digestive tracts and may not be able to eat enough to sustain themselves during lactation. Owners of such dogs may need to start feeding puppies supplemental food at an early age.

Senior dogs

Senior dogs experience decreased activity and slowed metabolism, and thus older dogs need 20 percent fewer total calories than do average adult dogs. Because of these changes, senior dogs tend to become overweight. Monitor your senior dog with the guidance of your veterinarian for changes in activity levels and weight, and adjust his diet appropriately to maintain proper body condition.

Diets

The number of food choices available to today's dog is truly astonishing! In the past few years, a multitude of extruded foods,[4] as well as refrigerated or frozen, freeze-dried, and other storable raw foods (dehydrated, pre-mixes, raw-coated), have been introduced to the pet food market. In addition to raw food diets, there are grain-free diets, and low carbohydrate diets. In the face of widely varying claims and even accusations about the relative merits of these different canine diets, determining what diet is right for your dog can be a complex process, and one that raises a number of questions. Note that, in the US, the AAFCO has established nutritional standards for dogs, and any commercial diet should clearly state on the packaging that the food has met or exceeded AAFCO nutrient profile recommendations. In the UK, The European Pet Food Industry Federation provides a Code of Good Labeling Practice. In the end, it comes down to caregiver preference, but in the next section we will explore these different types of diets, and offer information on some pros and cons of each.

Raw food diets

Raw food diets began showing up in homes in the 1990s with the invention of the BARF (biologically appropriate raw food or bones and raw food) diet, developed in Australia and thought to be superior due to the anti-aging benefits and nutritional superiority of uncooked meat.

---

[4] Extrusion is a process of grinding, mixing, and heating foods to be highly digestible. Kibble is an example of extruded food (as are many of your breakfast cereals).

Although the claims of anti-aging properties have yet to be documented scientifically, there are some benefits to raw diets.

Pros

1. Raw diets may feature more quality ingredients, including antioxidants, than those available in commercially prepared diets.
2. Concentration and types of protein may be more consistent.
3. Digestibility of raw protein can be higher or at least equal to that of commercially prepared foods.

Cons

1. Nutritional imbalance may occur in prepared or homemade raw diets (Freeman and Michel 2001). The greatest concern in formulating raw food diets is the proper calcium to phosphorus ratio. If diets are too high in muscle meat, phosphorus content can be too high. Calcium-rich foods need to be included, otherwise dogs are at risk of bone disorders. In a study of 6-week-old large breed puppies fed a bones and raw food (BARF) diet from about 3 weeks of age, nutritional osteodystrophy was reported in some of the litters (DeLay and Laing 2002). Nutritional secondary hyperparathyroidism has also been reported in a litter of German shepherd puppies fed a diet of 80 percent rice with 20 percent raw meat. The diet contained excessive amounts of phosphorus (Kawaguchi et al. 1993).
2. If bones are fed, there is a danger of perforation of the esophagus, stomach, or bowel (Rousseau et al. 2007).
3. Dogs are not as impervious to bacteria as once thought. Dogs can be susceptible to foodborne pathogens such as *Salmonella* that have been discovered in raw diets. One study reported isolation of *Salmonella* from 80 percent of homemade raw diets for dogs (Joffe and Schlesinger 2002). There is also a risk of fecal shedding of *Salmonella* because of potential concerns for people handling the food and bowls as well as handling feces from dogs fed raw diets. Thirty percent of fecal samples from dogs fed those diets also contained *Salmonella* (Joffe and Schlesinger 2002). In a more recent study of raw food diets, researchers found *Escherichia coli* in all 39 fecal samples (Nilsson 2015). In comparison to commercially prepared conventional diets versus raw diets, one study found that almost 6 percent of the raw food diets contained *Salmonella*, while none of the commercially prepared conventional diets contained this bacteria (Strohmeyer et al. 2006). Safe handling of all diets is essential to prevent transmission of disease to dogs and to humans handling the diet.

Grain-free and low carbohydrate diets

Dog foods often follow the same nutritional trends as human diets (Case 2014). Grain-free pet foods are currently very popular in human diets and are the new trend in dog foods. These foods were developed more in response to consumer demands rather than by any nutritional need of dogs, and whether or not these diets are actually healthier is still under debate. Claims of lower inflammation, reduced allergies, and other immune benefits lack support in the scientific literature (but many dog caregivers swear by their benefits).

Pros

1. Dogs with allergies to certain grains, including wheat, may show decreased symptoms from a grain free diet. Elimination diets prescribed by a veterinarian can help determine which foods cause reactions in your dog.
2. Caregivers report fewer immune issues and other health-related benefits such as decreased skin irritation, ear infections, and gastrointestinal upset (Billinghurst 1993).

Cons

1. Carbohydrate content may be higher in some grain-free foods.
2. There is a lack of scientific evidence to support the health claims of grain-free diets (Joffe and Schlesinger 2002).
3. Caregivers may feel that the ingredients in grain-free diets are more "natural," but dogs have evolved to be able to digest grains as well as many other sources of carbohydrates. Dogs are scavengers and omnivores that have adapted to a number of sources of carbohydrate.
4. Research suggests that grains are not a particularly common cause of pet allergies. In a review of the literature on dog food-related allergies, beef was found to be the most common allergen, with dairy as the second most frequent cause (Verlinden et al. 2006).

Feeding guidelines for your dog

Setting a feeding schedule

A general recommendation is that all dogs be fed twice daily. You can divide the amount of food your dog requires into two meals, spaced 8 to 12 hours apart, or break apart the dog's daily ration to be fed as treats throughout the day. Dogs are fed in a number of ways that meet both the owner's and the animal's needs. These methods include portion-control, free-choice and timed feeding.

Portion-control feeding refers to controlling the amount of food that your pet consumes by measuring your pet's food and providing it in one or more meals daily. This method is often used for weight control programs and for behavior modification programs. One simple trick to accomplish portion-control (particularly in a multi-dog-parent household, and/or when treats are regularly given to the dog) is to measure out your dog's daily food portion, including treats used for training or rewards, in the morning. Place this food in a single location, refrigerated if necessary. Any food given to the dog throughout the day is taken from this location, and any food remaining at the end of the day becomes your dog's dinner.

Free-choice feeding allows food to be available to your pet at all times, as much as your pet wants, and whenever he or she wants it. This method is best when feeding dry food, which will not spoil when left out. Most nursing mothers are often free-choice fed in order to provide them with the additional calories needed for lactation, but many dogs will overeat when fed in this manner, resulting in obesity. Free-choice feeding can lead to disinterested, slow eaters. In addition, free-choice feeding is not a good option for dogs with food-related behavioral issues such as resource guarding of the food bowl.

Timed feeding involves making a portion of food available for your pet to eat for a specific period of time. For example, food can be placed in the dog's bowl for 30 minutes. After that time, if the pet has not consumed the food, it is removed.

How much to feed?

It is best to think about caloric content rather than weight or volume when determining how much to feed your dog. Pet foods will have varying caloric content and variations in their feeding instructions. Do not assume that one cup of one diet equates to one cup of another; read the nutritional information provided by the manufacturer when switching to a new brand or diet. You can also use a dog food calculator (such as the one available at the link below) to help determine the amount of food for your dog. Keep in mind that you need to monitor your dog's body condition and may need to adjust his caloric intake as he reaches his desired condition goals.

Dog food calculator: http://www.dogfoodadvisor.com/dog-feeding-tips/dog-food-calculator

Consequences of obesity in dogs

Obesity is a major concern in dogs, as it is in humans.[5] Obesity is defined as an excess of body fat sufficient to contribute to disease, and is linked not only with disease but also a shortened lifespan (Laflamme 2006). In a cross-sectional study of the prevalence of overweight body condition and obesity in dogs from five veterinary practices in the UK, researchers found that 35.3 percent of dogs were classed as an ideal body shape, 38.9 percent were overweight, 20.4 percent were obese and 5.3 percent were underweight. Even moderately overweight (25 percent greater body weight) dogs are at greater risk for early death (Kealy et al. 2002). Obesity is associated with a number of diseases and reduced lifespan. In a study of diet on Labrador retrievers, median lifespan for dogs on a 25 percent restriction diet was 13 years, whereas the non-restricted diet dogs lived 11.2 years on average (Kealy et al. 2002). Dog factors that

> ### Box 8.3. Neutering and obesity
>
> - Neutered males and spayed females are at greater risk of being overweight or obese (for a detailed discussion of neutering, see Chapter 9).
>
> - In a study of over 21,000 dogs, 38% of all castrated male and spayed female adult dogs were overweight or obese (Lund et al. 2006). Dog caregivers may benefit from obesity prevention education at the time of the procedure.
>
> - Weight changes due to neutering can generally be easily managed through diet and exercise.

---

5 The Pet Food Manufacturers Association (PFMA) has tracked obesity from 2009 to 2014 and has found that little has improved in the pet obesity epidemic with 77 percent of vets believing that the obesity problem has worsened since 2009 (http://www.pfma.org.uk/_assets/docs/PFMA_WhitePaper_2014.pdf).

contribute to obesity include: genetic predisposition,[6] reproductive management, and dietary/exercise management from the caregiver (Bland et al. 2009). Neutering has been associated with increased weight gain and obesity in dogs (Box 8.3) and is discussed in detail in Chapter 9. The weight gain consequences of neutering can be counteracted through attention to diet and increased exercise.

Human factors greatly influence the occurrence of obesity in dogs. Risk factors of caregivers associated with obesity are owner age (with older owners more likely to have overweight dogs), hours of weekly exercise, frequency of snacks/treats, and personal income, with low-income households having the least awareness of health risks of obesity (Courcier et al. 2010). Other studies examining environmental risk factors agree that caregivers often fail to recognize obesity in their dogs and the influence that their own habits have on the health of their dogs. In a study of dog caregivers looking at diets of dog owners and how these relate to weight of their dogs, researchers found that in households where owners ate with a nutrient-rich but low calorie diet, dogs were kept at normal weights (Heuberger and Wakshlag 2010). Conversely, in a study of overweight dogs, 39 percent of owners of overweight dogs thought that their dog was at an acceptable weight (White et al. 2011).

As with humans, the best strategy for managing weight loss in our dogs is to reduce calories *and* increase exercise. Our companion dogs rely on us to make good choices for them, as in the vast majority of cases, we are the ones providing diet and exercise opportunities to our dogs. Using a diet formulated for weight loss is an option, but caregivers must be careful to note the calories and amounts of food recommended by whichever diet is selected. In some instances, it may be appropriate to adjust volume alone and not change to a specially-formulated therapeutic diet, if the pet can lose weight with modest caloric restriction and without feeding below the label guidelines for providing calories necessary for ideal weight. That process will ensure the pet receives adequate nutrients.

There is little to no evidence showing that any nutritional supplements aid in weight loss in dogs. There is one FDA-approved pharmaceutical that is currently available for the management of obesity in dogs. As with any medication, discuss options with your veterinarian, as it may not be appropriate for every dog.

Therapeutic weight loss formulas may not assist in weight loss any better than reducing the amount of your dog's current diet, particularly if label guidelines are not strictly adhered to. Diet foods may not be necessary unless advised by your veterinarian. In her book, *Dog Food Logic: Making smart decisions for your dog in an age of too many choices*, Linda P. Case recommends bypassing the diet food for your overweight dog, and instead decreasing the number of calories by reducing treats, slightly reducing the amount of food, and as always, increasing exercise. A study of weight loss formulated diets revealed a wide variation in the caloric content and the recommended caloric content (Linder and Freeman 2010). As with all commercially available diets, caregivers need to read the labels carefully, and to adjust the amount fed accordingly. You may inadvertently increase your dog's calories significantly by assuming that all dog foods' calorie contents by volume are similar. Watch your dog for changes, and adjust his intake gradually until

---

6 A recent study found that a fifth of Labrador Retrievers carry a genetic variation of a gene that predisposes them to weight gain (Raffan et al. 2016).

you receive the desired weight. You should continue walking your dog and providing exercise, because even though you reach your dog's desired weight, walking will continue to provide health benefits for you and your dog – as well as make a happier dog and a happier you!

Owner perceptions and obesity in dogs

As noted earlier, there is clear evidence that obesity in dogs is affected by the interrelationships between food management, exercise, and social factors. In a recent study of veterinarian and caregiver perspectives on obesity in dogs, veterinary practices felt that 3 percent of cases could be attributed to dog-specific factors and 97 percent to human specific factors such as diet, exercise, and owner beliefs (Bland et al. 2010). The way humans care for their dogs contributes to obesity. Bland and colleagues (2009) looked at human behavior's influences on obesity in dogs. They found that animals of normal weight were fed with two meals, whereas caregivers with obese dogs fed their dogs meals in either one portion or in three plus meals per day. Almost all caregivers in the study (99 percent) fed treats, but caregivers with normal weight dogs gave treats significantly less frequently than households with obese or mixed weight dogs. The frequency of exercise also differed between households, with normal weight dogs being exercised daily, compared to weekly for overweight or obese dogs. Recommended management strategies included reducing food intake, reducing treat feeding, and changing diet, before recommending more intensive options such as obesity clinics. Of the surveyed veterinary practices, 43 percent ran obesity clinics and 79 percent of those believed they were a valuable management tool. Of veterinary practices that did not run obesity clinics, only 46 percent believed them to be a valuable management tool. Dog owners generally preferred to try to reduce dog weight through diet manipulation, increasing exercise, and elimination of treats, prior to consulting the veterinary practice. Based on research projects like these, caregiver attitudes and behavior need to change in order to halt the increasing trend of obesity and overweight in dogs. Box 8.4 lists tools caregivers can use to overcome struggles with weight loss. Weight loss is not easy, for humans or for dogs. Our canine companions may present us with many challenges as we attempt to reduce their weight, but we must consider the negative health consequences of excess body weight and encourage healthy habits in our dogs.

Maintaining physical health

Physical health in your dog is dependent on diet, exercise, genetics, and a dose of good luck. To have a physically healthy dog, take him for walks, feed him well (but not in overabundance), and develop your relationship through shared activities. These few simple guidelines will not only help the physical well-being of your dog and help ensure a long life, but will also help keep your dog happy throughout that long life.

Box 8.4. Possible solutions to common issues impeding weight loss. (Adapted from Brooks et al. (2014). 2014 AAHA weight management guidelines for dogs and cats.)

- use food as salary the pet must earn;
- provide environmental enrichment;
- use food balls and food puzzles;
- place food to encourage exercise (hide/fetch);
- choose low-calorie treats (e.g., low-starch vegetables);
- remove pet from human feeding areas;
- explore separate meal feeding options;
- change food for all pets if possible;
- offer food puzzles to slow down and separate feedings;
- use treat allowance of up to 10% of the calories of the diet as a palatability enhancer;
- avoid offering alternatives if the pet skips a meal;
- consider water therapy/physical activity program;
- understand the difficulty of weight loss;
- find support groups;
- understand that food-seeking behavior is often attention-seeking behavior;
- understand that nutrient and calorie needs are met and that the begging is behavior, not nutritional or hunger-related;
- find other social or activity to substitute for food (e.g., play, groom, walk, offer affection);
- distribute a portion of the diet as treats instead of meals;
- divide food into more frequent, smaller meals.

Box 8.5. Websites for additional information on obesity in dogs

American Animal Hospital Association www.aahanet.org

Association for Pet Obesity Prevention www.PetObesityPrevention.org

Association of American Feed Control Officials (AAFCO) http://petfood.aafco.org/Labeling-Labeling-Requirements

Pet Food Manufacturers Association (PFMA) http://www.pfma.org.uk/_assets/docs/PFMA_WhitePaper_2014.pdf

Pet Nutrition Alliance www.petnutritionalliance.org/About_Us.aspx

Take-home messages

1. Evaluate your dog's diet and exercise requirements as an individual. Your dog's needs may differ from other dogs of the same breed, size, age, or type.
2. Start exercise programs slowly and increase duration and intensity over time (discuss any concerns with your veterinarian).
3. Daily exercise can improve your dog's health and well-being as well as your own.
4. Provide variety such as walks in new places and introduction of toys to keep exercise exciting.
5. Explore dog sports and activities to help foster exercise, health, and fun with your dog.
6. Monitor your dog's body condition and adjust their diet to assure their best health.
7. If you are feeding treats or scraps, adjust your dog's diet as necessary to control weight gain and obesity.
8. Keeping your dog lean and avoiding obesity is a great way to assure a longer life for your dog.

References and additional resources

American Pet Products Association (APPA). (2016). 2015–2016 APPA National Pet Owners Survey. Greenwich, CT: American Pet Products Association. Available at: http://www.americanpetproducts.org/pubs_survey.asp

ASPCA. (2016). Pet statistics: Facts about pet ownership in the U.S. (retrieved from http://www.aspca.org/animal-homelessness/shelter-intake-and-surrender/pet-statistics).

Association of American Feed Control Officials. (2011). *Official publication*. Oxford, Indiana: Association of American Feed Control Officials.

Baldwin, K., Bartges, J., Buffington, T. et al. (2010). 2010 AAHA Nutritional assessment guidelines for dogs and cats. *Journal of the American Animal Hospital Association* 46: 285–296.

Baumann, A., Russell, S., Furber, S., & Dobson, A. (2001). The epidemiology of dog walking: An unmet need for human and canine health. *The Medical Journal of Australia* 175: 632–634.

Billinghurst, I. (1993). *Give Your Dog a Bone: The practical commonsense way to feed dogs for a long healthy life*. Alexandria, Australia: Bridge Printery. 320 p.

Bjornvad, C.R., Nielsen, D.H., Armstrong, P.J., et al. (2011). Evaluation of a nine-point body condition scoring system in physically inactive pet cats. *American Journal of Veterinary Research* 74: 433–437.

Bland, I.M., Guthrie-Jones, A., Taylor, R.D., & Hill, J. (2009). Dog obesity: Owner attitudes and behaviour. *Preventive Veterinary Medicine* 92: 333–340.

Bland, I.M., Guthrie-Jones, A., & Taylor, R.D. (2010). Dog obesity: Veterinary practices' and owners' opinions on cause and management. *Preventive Veterinary Medicine* 94: 310–315.

Brooks, D., Churchill, J., Fein, K., et al. (2014). 2014 AAHA weight management guidelines for dogs and cats. *Journal of the American Animal Hospital Association* 50: 1–11.

Case, L.P. (2014). *Dog Food Logic: Making smart decisions for your dog in an age of too many choices*. Wenatchee, WA: Dogwise Publishing.

Christian, H.E., Westgarth, C., & Bauman, A. (2013). Dog ownership and physical activity: A review of the evidence. *Journal of Physical Activity & Health* 10: 750.

Cotman, C. W., & Berchtold, N.C. (2002). Exercise: A behavioral intervention to enhance brain health and plasticity. *Trends in Neurosciences* 25: 295–301.

Courcier, E. A., Thomson, R.M., Mellor, D.J., & Yam, P.S. (2010). An epidemiological study of environmental factors associated with canine obesity. *Journal of Small Animal Practice* 51: 362–367.

Currie, D. (2013). Dog walking often leads to better health, researchers find. *Nation's Health*. 42: 10.

Cutt, H., Giles-Corti, B., & Knuiman, M. (2008). Encouraging physical activity through dog walking: Why don't some owners walk with their dog? *Preventive Medicine* 46: 120–126.

Degeling, C., Burton, L., & McCormack, G.R. (2012). An investigation of the association between socio-demographic factors, dog-exercise requirements, and the amount of walking dogs receive. *Canadian Journal of Veterinary Research* 76: 235–240.

DeLay J., & Laing J. (2002). Nutritional osteodystrophy in puppies fed a BARF diet. *AHL Newsletter* 6: 23.

Dodman, H. (2009). *The Well-Adjusted Dog: Dr. Dodman's 7 steps to lifelong health and happiness for your best friend.* New York, NY: Houghton Mifflin Company.

Epping, J.N. (2010). Dog ownership and dog walking to promote physical activity and health in patients. *Current Sports Medicine Reports* 10: 224–227.

Freeman, L., & Michel, K.E. (2001). Evaluation of raw food diets for dogs. *Journal of the American Veterinary Medical Association* 218: 705–709.

Grandjean, D., Kelly, N.C., & Wills, J.M. (1996). Nutrition of racing and working dogs. In N. Kelly and J. Wills (Eds.). *BSAVA Manual of Companion Animal Nutrition & Feeding.* Ames, IA: Iowa State Press, pp. 63–92.

Heuberger, R., & Wakshlag, J. (2011). The relationship of feeding patterns and obesity in dogs. *Journal of Animal Physiology and Animal Nutrition* 95: 98–105.

Jeusette, I., Greco, D., Aquino, F., et al. (2009). Effect of breed on body composition and comparison between various methods to estimate body composition in dogs. *Research in Veterinary Science* 88: 227–232.

Joffe, D.J., & Schlesinger, D.P. (2002). Preliminary assessment of the risk of *Salmonella* infection in dogs fed raw chicken diets. *Canadian Veterinary Journal* 43: 441–442.

Kawaguchi, K., Braga, III I.S., Takahashi A., et al. (1993). Nutritional secondary hyperparathyroidism occurring in a strain of German shepherd puppies. *The Japanese Journal of Veterinary Research* 41: 89–96.

Kealy, R.D., Lawler, D.F., Ballam, J.M., et al. (2002). Effects of diet restriction on life span and age-related changes in dogs. *Journal of the American Veterinary Medical Association* 220: 1315–1320.

Kittleson, M.D., Keene, B., Pion, P.D., & Loyer, C.G. (1997). Results of the multicenter spaniel trial (MUST): taurine- and carnitine-responsive dilated cardiomyopathy in American cocker spaniels with decreased plasma taurine concentration. *Journal of Veterinary Internal Medicine* 11: 204–211.

Laflamme, D.P. (2006). Understanding and managing obesity in dogs and cats. *Veterinary Clinics Small Animal Practice* 36: 1283–1295.

Linder, D.E., & Freeman, L.M. (2010). Evaluation of calorie density and feeding directions for commercially available diets designed for weight loss in dogs and cats. *Journal of the American Veterinary Medical Association* 236: 74.

Lund, E., Armstrong, P., & Kirk, C. (2006). Prevalence and risk factors for obesity in adult dogs from private US veterinary practices. *International Journal of Applied Research in Veterinary Medicine* 4: 177–186.

Nap, R.C., Hazewinkel, H., Voorhout, G., et al. (1991). Growth and skeletal development in Great Dane pups fed different levels of protein intake. *The Journal of Nutrition* 121: S107–S1013.

National Research Council (U.S.). (2006). *Nutrient requirements of dogs and cats.* Washington, D.C: National Academies Press.

Newton, P.M., Mow, V.C., Gardner, T.R., et al. (1997). The effect of lifelong exercise on canine articular cartilage. *The American Journal of Sports Medicine* 25: 282–287.

Nilsson, O. (2015). Hygiene quality and presence of ESBL-producing Escherichia coli in raw food diets for dogs. *Infection Ecology & Epidemiology* 5: http://dx.doi.org/10.3402/iee.v5.28758

Raffan, E., Dennis, R.J., O'Donovan, C.J., et al. (2016). A deletion in the canine POMC gene is associated with weight and appetite in obesity-prone Labrador retriever dogs. *Cell Metabolism* 23: 893–900.

Richards, E.A., Troped, P.J., & Lim, E. (2014). Assessing the intensity of dog walking and impact on overall physical activity: A pilot study using accelerometry. *Open Journal of Preventative Medicine* 4: 523–528.

Richards, E.A., McDonough, M., Edwards, N., et al. (2013). Psychosocial and environmental factors associated with dog walking. *International Journal of Health Promotion and Education* 51: 198–211.

Rousseau, A., Prittie, J., Broussard, J.D., et al. (2007). Incidence and characterization of esophagitis following esophageal foreign body removal in dogs: 60 cases (1999–2003). *Journal of Veterinary Emergency and Critical Care* 17: 159–163.

Sanderson, S.L. (2006). Taurine and carnitine in canine cardiomyopathy. *Veterinary Clinics of North America: Small Animal Practice* 36: 1325–1343.

Sessa, W.C, Pritchard, K., Seyedi, N., Wang, J., & Hintze, T.H. (1994). Chronic exercise in dogs increases coronary vascular nitric oxide production and endothelial cell nitric oxide synthase gene expression. *Circulation Research* 74: 349–353.

Strohmeyer, R.A., Morley, P.S., Hyatt, D.R., et al. (2006). Evaluation of bacterial and protozoal contamination of commercially available raw meat diets for dogs. *Journal of the American Veterinary Medical Association* 228: 537–542.

The National Academies of Sciences, Engineering, and Medicine (2015). *Your Dog's Nutritional Needs: A science based guide for pet owners*. Washington DC: The National Academy of Sciences.

TNS research in association with PFMA, 2009 (2013). *(Nine in ten) Pet Obesity: The reality in 2009* – London Vet Show survey, Q. 8, 2013; *Pet Obesity: Five years on* – LM research in association with PFMA, Qs. 1 & 3, 2014.

Verlinden, A., Hesta, M., Millet, S., Janssens, G.P.J. (2006). Food Allergy in Dogs and Cats: A review. *Critical Reviews in Food Science and Nutrition* 46: 259–273.

Westgarth, C., Christley, R.M., & Christian, H.E. (2014). How might we increase physical activity through dog walking? A comprehensive review of dog walking correlates. *The International Journal of Behavioral Nutrition and Physical Activity* 11: 83-97.

White, G.A., Hobson-West, P., & Cobb, K. (2011). Canine obesity: Is there a difference between veterinarian and owner perception? *The Journal of Small Animal Practice* 52: 622–626.

World Small Animal Veterinary Association (WSAVA). (2011). Nutritional Assessment Guidelines. *Journal of Small Animal Practice* 52: 385–396.

Web resources

Agility Association of Canada (AAC). (2013). Official Website. Retrieved from http://aac.ca

American Kennel Club (AKC). (2012a). Official Website. Retrieved from http://www.akc.org/

Canadian Kennel Club (CKC). (2013). Official Website. Retrieved from http://www.ckc.ca/en/

International Federation of Cynological Sports (IFCS). (2002–2006). Official Website. Retrieved from http://www.dogsport.ru/

Integrative Veterinary Care: Feeding Large Breed Puppies
http://ivcjournal.com/feeding-large-breed-puppies/

US pet ownership estimates from the APPA for 2012: http://www.humanesociety.org/issues/pet_over population/facts/pet_ownership_statistics.html

PetMD: http://www.petmd.com/dog/centers/nutrition/evr_multi_what_is_grain_free_pet_food_really

Pet Food Industry.com: http://www.petfoodindustry.com/articles/4474-grain-free-gluten-free-ensuring-your-petfood-claims-stand-up-to-scrutiny?v=preview

The Pet Food Manufactures' Association: http://www.pfma.org.uk

The European Pet Food Federation: http://www.fediaf.org/self-regulation/labelling/

Dog Food Advisor: http://www.dogfoodadvisor.com/dog-feeding-tips/dog-ideal-weight/

Hills Pet: What to feed your pregnant and nursing dog: http://www.hillspet.com/en/us/dog-care/nutrition-feeding/what-to-feed-a-pregnant-dog

The Ohio State University Veterinary Medical Center Nutrition Support Service: http://vet.osu.edu/vmc/companion/our-services/nutrition-support-service

Chapter 9

Veterinary care for your happy dog

Photo: iStock

An integral part of taking on the responsibility for a canine companion is committing to the lifelong medical care of that animal. If we (and our dogs) are lucky, and our dogs' lives are healthy and accident-free, this can be a relatively simple process of puppy vaccinations and wellness visits, routine check-ups, and perhaps the occasional minor medical issue. Keeping our dogs happy and healthy through good nutrition and regular activity (both physical and mental) can help keep medical costs to a minimum. On the other hand, if our dog suffers from a serious disease or accident, this can become a daunting, expensive proposition. Bringing a dog into our home comes with the responsibility of ensuring that this new family member gets the veterinary care that they need, to the very best of our (financial) ability. Realistically, very few of us can afford to pay for unlimited medical treatments for our dog, should a very serious medical issue arise. Sometimes, difficult and even heartbreaking decisions must be made about what we can really afford. But, "to the best of our ability" means just this – we commit to paying what we can afford for necessary veterinary care. In some areas, programs exist to provide financial assistance to pet owners facing an unexpected medical crisis and who cannot afford a required treatment for their dogs. Many veterinary clinics support pet health insurance, to protect against unexpected medical expenses for your dog, and some practices will arrange monthly payment plans (versus payment in full due at time of service). However, if we cannot afford to provide even *basic* veterinary care, we should not take on the responsibility of caring for a dog. Costs of providing veterinary care vary according to geographic location; talk to dog-savvy friends and neighbors in your area about typical veterinary costs, or perhaps speak to a local veterinarian for an estimate of what the typical costs of care would be, per year, at least for the first year or two of your dog's life with you. A large nationwide study conducted by the American Veterinary Medical Association and Bayer Healthcare Animal Health in 2011 reported that visits to veterinarians had declined over the previous 10 years, with 51 percent of the 401 veterinary practices surveyed reporting fewer patients in the preceding two years (Volk et al. 2011). One of the most common reasons given by clients for not bringing their pets to the veterinarian was cost, with 53 percent of the pet owners surveyed stating that veterinary costs were much higher than expected (Volk et al. 2011). In some cases, inability or unwillingness to pay for treatment means an owner will request euthanasia for a dog with a treatable issue. In addition to the often tragic outcomes for the dogs in these cases, these situations are highly stressful for everyone: for owners, who have bonded with these dogs, and for veterinarians (and their staff), who have gone into their chosen profession specifically to protect the well-being of non-human animals. Cost of providing care should be considered before adopting a dog.

The goal of this chapter is to provide the most current, evidence-based recommendations available at the time of publication, for routine veterinary care throughout the life of your dog, vaccination schedules, and questions of reproductive health (to neuter or not to neuter? and, if so, when?). We will include some information on known breed predilections for certain diseases. We will cover tips on choosing a veterinarian, including the importance of consistency of care, and of working with a veterinarian who uses low-stress handling techniques (designed to minimize stress to your dog during necessary clinic visits). As with the rest of this book, this chapter is an overview, but we will provide references and additional resources that you can consult for more detailed information, should you wish to learn more.

An integral part of taking on the responsibility for a dog is committing to providing them with lifelong veterinary care, to the very best of our ability.

Recommended basic veterinary care schedule for the life of your dog

One of the first steps most owners take when they bring home a new puppy is to schedule a veterinary visit, during which the puppy is given a full physical exam, vaccinations may be given, and so on. And, most of us have received one of those cheerful "reminder" cards in the mail from our local veterinarian, letting us know that our pet is due for vaccinations, a physical exam, or similar, in the near future. It is helpful, however, to know in advance what to expect in terms

Figure 9.1: Providing basic and necessary veterinary care is part of the commitment made when bringing a dog into your home. (Photo: iStock)

of baseline[1] veterinary care, for the life of your dog. In 2012, the American Animal Hospital Association (AAHA) produced their *Canine Life Stage Guidelines*,[2] designed to help practicing veterinarians create comprehensive, individualized wellness plans for each stage of their canine patients' lives. The guidelines focus on providing comprehensive life stage wellness care as a means to improve quality of life and longevity of our dogs, via early detection and treatment or control of disease. Wellness care is preventative medicine, and can also result in significant cost savings in long-term healthcare expenses (Bartges et al. 2012).

One of the first recommendations is to choose a primary care veterinarian for your dog, and (once you have found a veterinarian that you like and respect) to see this same veterinarian consistently. Maintaining a relationship with your dog's veterinarian for the life of your dog ensures that your veterinarian will be familiar with your dog's medical and behavioral history, and allows for the best possible preventative care and treatment. Increasingly, veterinary specialists are called on to work on non-routine medical issues, and your primary care veterinarian will be the one to facilitate and coordinate this specialist care, should it be needed (Bartges et al. 2012). In some larger,

> *Your dog is an individual, and deserves individualized veterinary care.*

---

1 By "baseline" veterinary care, I'm referring to minimum regular veterinary care necessary for the prevention of disease and problems with parasites, early detection (leading to treatment) of any health problems, and other maintenance and wellness issues.
2 Available online at: https://www.aaha.org/professional/resources/canine_guidelines_abstract.aspx

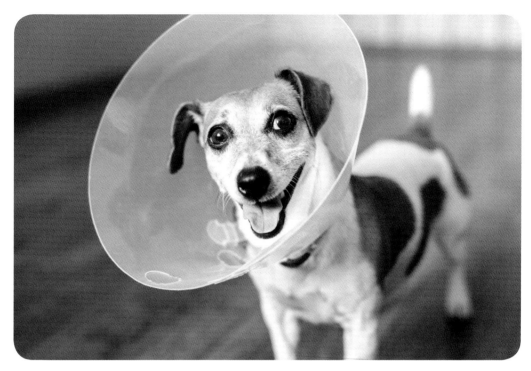

Figure 9.2: Veterinary care is a lifelong issue, but recommendations for baseline veterinary care will vary with the life stage of your dog. (Photo: iStock)

multiclinic practices, you may have to insist on seeing the same clinician each time you book an appointment; otherwise, some clinics will assign you to whichever clinician is on duty for that day. Anecdotally, I have always found that working with a familiar veterinarian can be less stressful for shy or fearful dogs, *if* that veterinarian is using low-stress handling protocols (more on these later in this chapter). Incidentally, in a recent study we carried out on stress responses to veterinary care in cats, familiarity with the veterinarian was found to play a primary role in reducing stress (Nibblett et al. 2014).

The next recommendations of the guidelines involve ensuring safe transport of the dog to and from the veterinary clinic (always use a leash, harness, or carrier), and the use of low-stress handling and restraint techniques in the clinic (more on these later). If your dog has a history of being very fearful and/or aggressive at the clinic, talk to your veterinarian about a fast-acting anti-anxiety medication[3] to be taken just prior to the visit, to help the experience be safer and less stressful for all concerned.

Generally, recommendations for necessary veterinary care are organized by life stage of your dog. Life stages for dogs are defined not just by age but by characteristic (e.g., puppy, adult, senior). Different sources define life stages in different ways, and the chronological ages encompassed by these life stages will vary with the breed and size (large vs. small) of the dog. In

---

3 More information on these medications can be found in Chapter 7; note that acepromazine, commonly used for this purpose in the past, is no longer recommended for use in this way. More information on the use (and misuse) of acepromazine can also be found on Debbie Jacob's wonderful site "FearfulDogs.com": http://fearfuldogs.com/acepromazine/

Box 9.1. Life stages of dogs

1. *Puppy*: From birth until the dog is reproductively mature; some authors will also distinguish the "socialization phase," given its importance to behavioral development (usually from 3 to 14 weeks of age).

2. *Adolescent* (aka "teenager" or junior): Reproductively mature, but still growing. Behaviorally, these dogs often behave like, well, teenagers: they are often very playful, curious, and energetic, and may be overly boisterous; they are still learning what is socially acceptable!

3. *Adult*: The dog has finished growing, and is now physically and socially mature. Some authors (particularly when discussing behavior) will separate physical and social maturity into two life stages, but social maturity in particular can vary widely between dogs and be difficult to identify conclusively. The "adult" stage takes a dog from social maturity to about 7 years of age.

4. *Senior*: The last 25 percent of the dog's expected lifespan. For many dogs, this means aged about 7 years or older.

5. *Geriatric*: The years beyond the dog's expected lifespan. An important reason for distinguishing this life stage from "senior" is the increased risk for certain health issues, such as cognitive dysfunction, during this stage.

For example, let's say I adopt a puppy with a life expectancy (based on breed) of 10 years. Puppyhood would run from birth to whenever the dog reached sexual maturity (usually between 6 and 12 months of age); adolescence usually runs up to between 18 months and 2 years, after which my dog is considered an adult, until he surpasses 7.5 years (after which he is considered a senior citizen). If he lives past 10 years, he could be considered geriatric.

Adapted from:

Bartges et al. (2012). AAHA canine life stage guidelines. *Journal of the American Animal Hospital Association* 48: 1–11.

Martin, K.M., and Martin, D. (2011). *Puppy Start Right: Foundation training for the companion dog.* Waltham, MA: Karen Pryor Clickertraining. 184 p.

---

general, life expectancy of domestic dogs ranges between 8 and 12 years, although small breed dogs have a longer life expectancy than large breed dogs. Your veterinarian can advise you here, as can a reputable, knowledgeable breeder or breed association for your purebred dog. Box 9.1 summarizes one way in which canine life stages are classified. Box 9.2 provides a summary of the recommendations for optimal veterinary wellness care at each of these life stages, based largely on the AAHA Guidelines.

Protecting your dog from disease: the importance of vaccinations

The science of vaccination is not new; vaccines today are the product of many, many years of research, by a countless number of scientists and researchers, who continually evaluate and re-evaluate both the benefits and the risks of vaccines, and persistently strive to improve their

Box 9.2. Wellness recommendations for baseline veterinary care for your dog

These are items which should be completed by your veterinarian at each of your dog's life stages, or should be discussed with your vet at each stage, if you have any questions or concerns. This is not meant to be an exhaustive list of everything your vet may address, *or* that your dog may need at some point in his life. Always work with your veterinarian to safeguard and maintain the health of your canine companion.

For all life stages:

Regular (yearly is recommended) full physical exams;

Oral exam to check teeth (with dental cleanings performed promptly, when necessary);

Evaluate need for core and non-core vaccinations[a] or titers, and vaccinate if necessary;

Assess body condition[b] of your dog and adjust diet and/or exercise if required;

Practice year-round (or seasonal, depending on climate) ecto- and endoparasite[c] control (fleas, ticks, worms);

Address any questions or concerns about behavior[d] promptly, referring to a qualified behaviorist when necessary.

Additionally, for puppies:

Administer core vaccinations (recommended for all puppies, and all dogs with an unknown vaccination history);

Follow early deworming guidelines[c], with regular (2–4) fecal exams during the first year of life;

Discuss neutering[e] risks and benefits. For dogs remaining intact, discuss genetic screening and counseling, and breeding plans;

Discuss any breed-specific health risks, and ways to minimize these risks;

Evaluate socialization and training plan; get recommendations if necessary for an experienced, reputable, force-free trainer and puppy class in your area.

Additionally, for adolescents and adults:

If your dog is a working or service dog, assess any special needs that she has and ensure these are being met.

Additionally, for seniors[f] and geriatric dogs:

Evaluate mobility, sight, and hearing, and discuss environmental modifications necessary to accommodate any changes with age;

Maintain a written record of your dog's cognitive capabilities[g], and consider diet and environmental enrichment changes to slow cognitive decline.

Adapted from:

Bartges et al. (2012). AAHA canine life stage guidelines. *Journal of the American Animal Hospital Association* 48: 1–11.

Hammerle, et al. (2015). AAHA canine and feline behavior management guidelines. *Journal of the American Animal Hospital Association* 51: 205–221.

Notes:

[a] Current recommendations concerning vaccinations will be discussed in a separate section, later in this chapter. Non-core vaccinations are optional vaccines that should be considered in light of the exposure risk of the animal, that is, based on geographic distribution and the lifestyle of the pet. Examples of these include distemper–measles combination vaccine and Bordetella bronchiseptica. Core vaccines include vaccines for canine parvovirus (CPV), canine distemper virus (CDV), canine adenovirus (CAV), and rabies (University of California at Davis VMTH Canine and Feline Vaccination Guidelines, available at www.vetmed.ucdavis.edu/vmth/small_animal/internal_medicine/newsletters/vaccination_protocols.cfm).
[b] Body condition scoring is discussed in detail in Chapter 8, on physical wellness.
[c] Based on Companion Animal Parasite Council and Centers for Disease Control (CDC) guidelines, available at www.capcvet.org/capc-recommendations/
[d] For more on the importance of behavioral wellness, particularly socialization of puppies, see Chapter 6.
[e] Current recommendations regarding neutering your pet, and age at neutering, will be discussed in a separate section, later in this chapter.
[f] AAHA also has a set of guidelines for senior dog care, available at https://www.aaha.org/professional/resources/senior_care.aspx
[g] Effects of aging on canine cognitive function are discussed further in Chapter 10 of this book.

efficacy and safety. In recent years, a number of task forces, composed primarily of expert veterinarians, scientists, and epidemiologists, have gotten together to make recommendations on the use of vaccines in companion animals.[4] Vaccines are one of the best ways to protect our dogs from disease: they provide life-saving benefits, are associated with minimal risk, and should be part of routine preventative care (Wellborn et al. 2011). Veterinarians today are advised to consider the cost–benefit ratio for each vaccine they give to their patients; Box 9.3 lists some approaches used to minimize the risks of vaccines, while still making use of their health benefits.

- *Core vaccines* are recommended for *all* puppies, and dogs with an unknown vaccination history. The diseases that these vaccines were developed to prevent are nothing to take lightly: they are widely distributed, common in many areas, are often readily transmittable, and cause significant disease and/or death. In addition, safe and effective vaccines are available for these diseases, that either provide immunity (i.e., prevent infection from occurring) or confer a high degree of protection from the disease (i.e., do not prevent infection, but may provide protection such that the dog will develop minimal if any clinical signs of the disease; Davis-Wurzler 2014). Core vaccines in the US include vaccines for canine parvovirus (CPV),

---

4 For example, the AAHA Canine Vaccine Task Force (https://www.aaha.org/professional/resources/canine_vaccine.aspx), the AVMA Council on Biologic and Therapeutic Agents (https://www.avma.org/About/Governance/Councils/Pages/Council-on-Biologic-and-Therapeutic-Agents-Entity-Description.aspx), and (for cats), the AAFP/AFM Advisory Panel on Feline Vaccines (http://www.catvets.com/guidelines/practice-guidelines/feline-vaccination-guidelines).

Box 9.3. Vaccination schedules in practice: recommendations for minimizing risks while still protecting your dog.

See also Box 9.2 on health care recommendations for the life of your dog

> Vaccines are to be used with forethought based on the risk of disease to the population and the individual, balanced with assessment of the risks associated with individual vaccines ... The goal is to reach the highest level of overall animal health with the minimum number of adverse events, based on scientific and epidemiologic merit
>
> (Davis-Wurzler 2014).

Health status: Dogs who are too young, too sick, prone to severe vaccine reactions, suffering from immune-mediated diseases, or not exposed to the disease in question, should not be vaccinated[2] (although dogs in the first two categories may be vaccinated once they are older and/or have physically recovered).

Individualized medicine: Types of vaccines recommended, and the frequency of vaccination, will vary depending on the lifestyle of the pet being vaccinated (e.g., geographic location, time spent with other dogs, travel or boarding plans, and so on). In addition, these factors may change over time. The vaccination plan for each individual dog should be decided by the owner at routine annual examinations, following a discussion between the veterinarian and the client regarding the animal's lifestyle in the year ahead.[3]

Vaccine intervals: Although it is medically acceptable to administer multiple vaccines at the same time,[1] some veterinarians will give puppy vaccines individually, separated by 10 days or so between injections, both to decrease chances of an adverse reaction, and to help identify which particular vaccine causes this reaction.[2] If the same product is being given serially during the course of puppy vaccinations, the interval should never be less than two to three weeks, to avoid any possible reductions in the vaccine's effectiveness.[1] Serial vaccinations are used to help ensure immunity in that puppy.[1] For adult dogs, the interval between different vaccinations should also be at least two to three weeks, to avoid interference between the two vaccines.[1]

Use of titers: Vaccine titers are blood tests that measure the concentration of antibodies (proteins produced by the immune system to combine with and destroy antigens, or foreign molecules, which enter the body) present in the dog's bloodstream. Although not a perfect measure of immunity, titers are useful in determining whether that dog still has immunity to the disease in question[2]. Instead of simply re-vaccinating on a preset schedule, many veterinarians will now use titers to assess whether another vaccination is needed at that time. Discuss the possible use of titers with your veterinarian, to help limit the number of vaccinations given to your dog to only those that are most beneficial to your dog.[4]

Sources:

1 Davis-Wurzler, G. (2014). 2013 update on current vaccination strategies in puppies and kittens. Veterinary Clinics of North America: Small Animal Practice 44: 235–263.
2 Royal, B. Choosing a Vaccination Schedule for Your Dog. TheBark.com (accessed May 2016).
3 University of California at Davis (2012). VMTH Canine and Feline Vaccination Guidelines. http://www.vetmed.ucdavis.edu/vmth/small_animal/internal_medicine/newsletters/vaccination_protocols.cfm (accessed May 2016).
4 Note that laws regarding acceptability of titers in lieu of vaccination for rabies vary by region; your veterinarian can provide information on laws in your area.

canine distemper virus (CDV), canine adenovirus (CAV), and rabies. In most areas, vaccination for rabies is required by law, given the highly infectious, zoonotic[5] nature of the disease.

- *Non-core vaccines* are optional vaccines that should be discussed with your veterinarian, in light of the exposure risk of your dog, based on where you live and the lifestyle of your dog (e.g., does your dog regularly go to the dog park and mix with unfamiliar dogs? will your dog be boarded soon? and so on …). Your dog may or may not need one or more of these vaccines, so discuss the relative risks and benefits with your veterinarian. Several of the diseases prevented by non-core vaccines are often self-limiting, or respond readily to treatment (UC Davis 2012). Vaccines considered as non-core vaccines are the vaccines against canine parainfluenza virus (CPiV), canine influenza virus, distemper–measles combination vaccine, *Bordetella bronchiseptica* (aka "kennel cough"), *Leptospira* spp. (leptospirosis, a serious, zoonotic bacterial infection), and *Borrelia burgdorferi* (aka "Lyme disease"). Note that vaccination with these vaccines may be less effective in protecting against disease than vaccination with the core vaccines (UC Davis 2012). In some cases, these vaccines may not give 100 percent protection against catching the disease, but should lessen the severity of clinical signs if the dog does contract that disease. The increasing focus on treating patients as individuals, and evaluating the risks and benefits to the individual dog (as well as the population in which that dog lives), is what has led to the differentiation between core and non-core vaccines (Davis-Wurzler 2014). Box 9.4 describes the three primary types of vaccines that may be given to your dog.

> *It is far better to prevent than experience disease.*
>
> (G.M. Davis-Wurzler, 2014)

The first 16 weeks: timing of puppy vaccines and exposure to the Big Wide World.

In the US, core vaccinations for puppies begin at 6 to 9 weeks of age, and continue (usually every three to four weeks) until the puppy is 4 months (16 weeks) old. Vaccine schedules in other countries may vary somewhat. In the past, dog owners were instructed to keep their puppies isolated until the complete series of puppy vaccinations was completed (i.e., until the puppies were 4 months old), in order to minimize risk of the puppy acquiring a communicable disease. However, while core vaccinations of puppies are still considered essential, isolation for the first 4 months of their lives is no longer considered essential or even advisable. Early socialization of puppies is very important to their lifelong behavioral well-being, and numerous publications now support the importance of exposing your puppy to many new people, places, and other (healthy) dogs, prior to 4 months of age (e.g., Hunthausen 2010; Stepita et

---

5 Zoonotic refers to diseases that can be transmitted from non-human animals to humans.

> **Box 9.4. Three primary types of vaccines used in veterinary medicine today**
>
> *Modified live vaccine*: Vaccines created by altering the pathogen in some way, so that it is no longer able to cause serious or clinical disease.
>
> *Killed vaccine*: Vaccines produced by inactivating the pathogen completely, making it incapable of reproducing and therefore unable to cause disease.
>
> *Recombinant vaccines*: There are multiple forms of this type of vaccine, which are produced using recombinant DNA technology. For example, in one form, DNA encoding an antigen (such as a bacterial surface protein) that stimulates an immune response is inserted into bacterial or mammalian cells, expressing the antigen in these cells and then purifying it from them.
>
> Some researchers categorize vaccines into two groups: infectious (meaning having the potential, under certain relatively rare circumstances, to cause disease in some individuals, with immune-compromised individuals being at highest risk), and non-infectious (which lack the ability to reproduce within a host individual, and therefore cannot cause disease). Modified live vaccines fall under the "infectious" category, while killed and recombinant vaccines fall under the "non-infectious" category.
>
> *Sources:*
>
> Davis-Wurzler, G. (2014). Update on current vaccination strategies in puppies and kittens. *Veterinary Clinics of North America: Small Animal Practice* 44: 235A263.
>
> Nature.com. Recombinant vaccine subject page http://www.nature.com/subjects/recombinant-vaccine (accessed May 2016).

al. 2013). Puppies born to healthy, vaccinated mothers, who are receiving their regularly-scheduled puppy vaccines, and who are socialized sensibly (with other healthy dogs, for example, and in areas free from high risk of environmental contamination), have a low risk of contracting infectious disease (Stepita et al. 2013). The risks of behavioral problems commonly associated with insufficient socialization are far, far greater than the infectious disease risks. So, ensure that your puppy is healthy, and receiving its recommended vaccinations, and exercise common-sense precautions when socializing your puppy. For example, enroll in a reputable puppy class in which all puppies have been required to document health and vaccination status prior to joining the class, and that is held in a clean (and cleanable) environment (parvovirus can survive in the environment for up to two years). That big, busy, off-leash dog park full of unfamiliar dogs is not the place for your new puppy.

> *There is no medical reason to delay puppy or kitten classes or social exposure until the vaccination series is completed as long as exposure to sick animals is prohibited, basic hygiene is practiced, and diets are high quality.*
>
> (Hammerle et al. 2015, p. 32)

Reproductive health: To neuter or not to neuter?

For most dog caretakers and owners who do not have any plans to breed their dogs, the question of whether or not to neuter[6] comes up at some point.[7] Until quite recently, dog owners (in the US, at least) were generally advised to neuter their dogs, often at a very young age.[8] Owners were told that benefits of neutering (including reduced risk of certain cancers, behavioral benefits, and ethical issues involving euthanasia of unwanted dogs) outweigh the risks associated with these very common surgeries. In recent years, however, the decision has become a bit more controversial, as more research comes to light on the long-term health effects of neutering. Even if an owner decides to neuter his dog, the question of *when* to neuter ("pediatric" neutering? before one year? or, only after one year?) has also been added to the picture. Many owners are now faced with conflicting advice, depending on whether that advice comes from their veterinarian, the Internet, a breeder, or their friends. So, what now? What is the best course of action, when it comes to neutering your dog? The answer is (as it so often is in life), "it depends." Many of the risks and benefits of neutering vary with sex of the dog involved. The age at which the dog is neutered also plays a role in some of these issues, as does the breed of the dog (for example, some breeds are predisposed to certain types of cancer or joint diseases that are rarely seen in other breeds). The research is ongoing, and we still have more to learn. This section will summarize the pros and cons of neutering, based on the science and expert recommendations available to date. The "References and additional resources" section at the end of this chapter contains a list of the original research papers cited in this summary, should you wish to read in more detail about any of these issues.

It is important to note, however, that on two issues, the vast majority of experts in veterinary medicine and canine welfare agree:

1. Most of these risk/benefit discussions surrounding neutering are aimed at *"owned" dogs*, that is, dogs living in a home with human caretakers committed to their well-being. In the case of unowned dogs such as those in shelter situations, the benefits of neutering greatly outweigh the risks, from an ethical and population perspective. Over a million dogs are euthanized *each year* in shelters in the US, possibly multiple millions, depending on which source you read; and only around 10 percent of the animals received by shelters have been spayed or neutered prior to admittance (ASPCA,[9] National Council of Pet Population Study and Policy 1994). Only about 26 percent of dogs entering the shelter as strays are successfully reunited with their human families (ASPCA), adding more weight to the recommendations

---

6 In this chapter, we will use "neuter" to refer to alteration of male or female dogs to prevent reproduction. When referring to a surgery done only on one sex (e.g., spay for female dogs, or castration for male dogs), we will do so by name. There are a number of different surgeries done for neutering purposes; your veterinarian can explain which surgical procedure he/she would be using.
7 These discussions, of course, are less relevant to professional breeders, although they may be important information for purchasers of puppies to consider when planning for life with their new canine companion.
8 Note that this is not the case in many European countries, where neutering is not routinely recommended (Hart et al. 2014).
9 ASPCA Shelter Intake and Surrender Pet Statistics page, *http://www.aspca.org/animal-homelessness/shelter-intake-and-surrender/pet-statistics* (accessed May 2016).

(or requirements, in the UK and Ireland) to microchip your dog, and ensure that she is wearing identification when outside the home! The number of stray dogs living in the US is unknown, but is likely in the millions (estimates for the stray cat population range up to 70 million; ASPCA). So, the recommendation to neuter dogs living in shelter situations (or adopted from them) remains firm.

2. For companion dogs, living in a home with committed human caretakers, the decision on whether to neuter and when to neuter should be made on an individual basis, with the primary concern being the lifelong health of that individual animal. This decision should take into account risks of certain types of medical issues (how common these issues are, and how treatable if contracted) and behavioral issues (how serious the issues may be, and how ready and willing the human household is to manage an intact dog). Discussions should take into account the breed, size, background, and lifestyle of the dog, as these can all influence relative risks and benefits.

Neutering and cancer: risks and benefits – consider breed and sex (and age) of your dog

Hormones associated with the reproductive organs influence many tissues in the body, and have been associated with development of some types of cancers (Smith 2014). Thus, neutering can play a role in cancer risks. For some types of cancer, neutering appears to reduce or even eliminate risks; for others, it may actually increase risks. Risks of certain types of cancers are very low for some breeds, but higher in others, and effects of neutering on cancer risk also appear to vary with breed (at least for some breeds), so considering breed in evaluating relative risks and benefits of neutering is important and supported by the science. In some cases, waiting to neuter until after one year can reduce the risks associated with neutering (for example, in the case of increased risk of osteosarcoma in neutered dogs). In others, waiting to neuter can reduce the benefits associated with neutering (e.g., the reduced risk of mammary tumors in neutered dogs). How you weigh these factors depends on the risks in your own dog, and these risks depend (at least in part) on breed, size, and where you live: how likely is your dog, given these factors, to contract any of these diseases? Your veterinarian and/or local breed organization can provide helpful information on these questions.

The most commonly cited benefit of spaying is a significant reduction in the risk of *mammary tumors*, which are commonly malignant and highly metastatic (i.e., will spread throughout the body). Mammary tumors are the most common form of tumors in intact female dogs, with one study noting that mammary tumors comprised 53.3 percent of all malignant tumors in these dogs, with highest risks in boxers, cocker spaniels, English springer spaniels, and dachshunds (Moe 2001). Overall incidence of mammary tumors in dogs is 3.4 percent, with about 50 percent of these being malignant (numerous studies, summarized in Root Kustritz 2014). Risks increase with age and exposure to female sex hormones (Egenvall et al. 2005; Dobson et al. 2002); mean age of development of mammary tumors in intact female dogs is around 9 to 11 years (although this will vary with breed; Smith 2014; Moe 2001). This risk may be largely eliminated by spaying before the dog's first estrus cycle, to a lifetime risk[10] of 0.5 percent; if spayed between the first and second estrus cycle, lifetime risk increases to 8 percent, and after the second estrus cycle, to about 26 percent (Schneider et al. 1969; Smith 2014). One study reported that intact dogs have 7 times the risk of developing mammary tumors when they get older, compared to the risk of neutered dogs (Dorn et al. 1968). It should be noted, however, that not all researchers are convinced of the protective effects of spaying against mammary tumors; in a 2012 review of the literature, Beauvais and colleagues concluded that the evidence for the link was not as sound as previously believed (Beauvais et al. 2012a). This lack of conclusiveness may be due in part to the different risks and benefits by breed and sex of the dogs involved in the various studies. Notable differences in cancer risks by breed and sex have been documented by various authors. For example, another recent study comparing effects of neutering in two breeds, Labrador retrievers and golden retrievers, found that these effects varied by breed and sex, with cancer risks for female golden retrievers actually increasing with neutering (this was not found for the male golden retrievers, or Labradors of either sex; Hart et al. 2014). Once again, this supports the recommendation to discuss this decision with your veterinarian; benefits and risks will vary by breed and sex, as well as age at which the surgery is done, and your veterinarian is the best source of information on this question. Obviously, the risk of cancer of the uterus, ovaries, or testes will be eliminated by surgical removal of these organs (ovariohysterectomy for uterine and ovarian neoplasia, and castration for testicular tumors; Smith 2014), although the risks for these types of cancers in domestic dogs are generally very low ($\leq 0.5\%$ in females; $< 1\%$ risk of dying of testicular cancer for males; summarized in Sanborn 2007).

On the "risks" side of the equation, a commonly-cited cancer-related risk of spay/neuter surgery are the increased risks of *osteosarcoma* (bone cancer) and *hemangiosarcoma* (cancers arising from the lining of the blood vessels, and commonly occurring in the heart or spleen). Both of these cancers are aggressive and tend to metastasize (i.e., spread to other parts of the body). However, the overall risk of these cancers in the domestic dog is low: 0.2 percent incidence for osteosarcoma (Root Kustritz 2010) and the same for hemangiosarcoma (Reichler 2009). In general, risk of osteosarcoma increases 1.3 to 2-fold with neutering (Priester and McKay 1980; Ru et al. 1998). Risk of hemangiosarcoma increases by a factor of 2.2 for splenic, and by a factor of 4 to 5 for cardiac, hemangiosarcoma in spayed females, and by a factor of 2.4 overall for castrated males (Prymak et al. 1985; Ware and Hopper 1999; Root Kustritz 2014). Again, breed can play a role in evaluating risk: a study in Rottweilers, a breed with higher risks of osteosarcoma, found that dogs of this breed neutered before 1 year of age were 3.1 to 3.8 times more likely to develop the disease than intact dogs (Cooley et al. 2002). If your dog is a female golden retriever, the relative increase in cancer risks associated with spaying may be greater (up to 3 to 4 times the level of intact females), based on two large studies done at the University of California in Davis, California (Hart et al. 2014; Torres de la Riva et al. 2013).[11] These last findings led the authors to suggest that, for female golden retrievers at least, the presence of female reproductive hormones may have a protective effect against cancer over the life of these dogs (Hart et al. 2014).

---

10 Lifetime risk is the risk of developing a disease at any point during one's lifetime.
11 This marked increased risk was not seen, for the most part, in male golden retrievers, or male or female Labrador retrievers.

Also on the "risks" side, neutering male dogs has been found to be associated with an approximately 4-fold increase in risk of *prostate cancer* compared with intact dogs (Teske et al. 2002; Sorenmo et al. 2003). Prostate cancer, however, is very uncommon in dogs (< 0.6%; Weaver 1981). There is some evidence that neutering may increase risk of *mast cell tumors* (White et al. 2011), particularly in certain breeds, as in female golden retrievers (Torres de la Riva et al. 2013; male goldens did not show this trend) and male and female vizslas (Zink et al. 2014). Relative risks increased 2 to > 4-fold following neutering (summarized in Smith 2014). It should be noted, however, that some authors have pointed out that neutering is associated with longer life span in dogs (perhaps due to some health benefits associated with neutering, or to the overall better care that owned, neutered dogs receive; Reichler 2009). Therefore, some of these reported increases in incidence of cancers associated with neutering may be due in part to the greater longevity in the neutered dogs (Root Kustritz 2014), as risks of many cancers increase with age.

Neutering and joint disorders: consider breed tendencies, and keep your dog fit

A number of increased risks for joint problems have been associated with neutering, particularly when the surgery is done before the dog has finished growing. The two most commonly reported issues are an increase in *cranial cruciate ligament rupture* and in the incidence of *hip dysplasia*.

The cranial cruciate ligament (also called the anterior cruciate ligament, or ACL) is the band of connective tissue that connects the thigh bone (femur) with the lower leg bone (shin or tibia/fibula), and helps to stabilize the stifle joint (or what would be called the knee, in humans). ACL rupture is the most common cause of rear-leg lameness in dogs, and can contribute to degenerative joint disease in the affected joint. Neutering is associated with a two-fold increase in the risk of ACL rupture (Whitehair et al. 1993), and increased risks are seen even when the effects of weight are taken into account (for, as we all know, carrying excess weight puts strain on our joints; Torres de la Riva et al. 2013; Duval et al. 1999). Although the reasons for this relationship are not fully understood, it has been suggested that, due to the fact that the growth and development of the skeletal system is controlled in part by reproductive hormones, neutering prior to the age when growth has completed may result in changes to the structure and biomechanics of the skeleton (Salmeri et al. 1991, and others; summarized in Root Kustritz 2007). Neutering immature dogs delays the closure of the growth plates in bones that are still growing, resulting in those bones being significantly longer than those same bones in intact dogs, or dogs neutered after maturity (Salmeri et al. 1991). Neutering carried out after some growth plates have closed, but while others are still growing, may result in alignment or stability problems in that dog's skeleton, which in turn could impact the dog's performance and the durability of his joints (Sanborn 2007).

These changes in joint confirmation may also contribute to the increased risk of hip dysplasia in neutered dogs. Hip dysplasia is an inherited disorder, is more common in larger dogs (but can be seen in dogs of any size), and can be controlled to some extent by diet. Incidence of hip dysplasia in domestic dogs is about 1.7 percent, with higher incidence in large- and giant-breed dogs such as German shepherd dogs, golden retrievers, Labrador retrievers, and Saint Bernards (among others; Root Kustritz 2007). A number of studies have reported an increase in hip dysplasia in dogs neutered at a young age (Hart et al. 2014; Spain et al. 2004). For example, Hart et al. (2014) reported that neutering at < 6 months of age markedly and significantly increased the occurrence of joint disorders in Labrador retrievers (the rate was double that of intact dogs) and golden retrievers (the rate was 4 to 5 times higher in the neutered dogs).

Neutering your dog: other issues

Pyometra

Pyometra is an acute infection of the uterus that affects about 23–25 percent of intact female dogs by the age of 10 years (Hagman et al. 2011), with Bernese mountain dogs, Rottweilers, rough-haired collies, Cavalier King Charles spaniels, and golden retrievers among the breeds having higher risk for pyometra, as well as intact females who have not whelped puppies (Niskanen and Thrusfield 1998). Spaying (ovariohysterectory) virtually eliminates the risk of pyometra. This surgery can be curative if done at the time of diagnosis in an intact dog; however, mortality when surgery is used to treat pyometra ranges from 0 to 17% (Johnston et al. 2001). In her review of the cost-benefits of neutering, Sanborn (2007) estimated that pyometra kills about 1 percent of intact female dogs.

Obesity

Changes in metabolism following neutering can contribute to *obesity* in many dogs. A number of studies have documented higher risks of obesity for neutered dogs: 2 times the risk of obesity in spayed versus intact females (Edney and Smith 1986); spayed females were 1.6 times more likely to be obese than intact females, and neutered males were 3 times more likely to be obese than intact males (McGreevy et al. 2005). This risk may be reduced for dogs neutered early (i.e., before 5.5 months of age), when compared to dogs neutered later in life (Spain et al. 2004). Weight, however, can be controlled to a large degree (and associated risks reduced) through diet and regular exercise.

Urinary incontinence in female dogs

Another commonly reported risk of spaying female dogs is *urinary incontinence*, due to sphincter mechanism incompetence and also referred to as hormonal urinary incompetence (Beauvais et al. 2012b). This phenomenon can occur almost immediately after the spay surgery is done, or can occur after a delay of several years, and the incidence rate in spayed females has been reported as between 4–20 percent (depending on study, and in contrast to 0.3% in intact females; Thrusfield et al. 1998; Stöcklin-Gautschi et al. 2001; Arnold et al. 1989). Risks have historically been considered higher for females neutered prior to their first estrus cycle, or before 3 months of age; however, recent work has not found this association between age of neutering and risk of urinary incontinence (Spain et al. 2004; Forsee et al. 2013). Moreover, some authors have questioned the strength of the evidence for a link between neutering and urinary incontinence (Beauvais et al. 2012b). If it occurs, urinary incontinence can be understandably challenging for the human household members (de Bleser et al. 2011), but generally be treated using hormonal supplements or medications, although this therapy may need to be lifelong (Shiel et al. 2008).

Cognitive function and impairment with age

There is some evidence that neutering may be associated with increased risk of progression of cognitive impairment in geriatric dogs (aka *cognitive decline and dysfunction*), perhaps due to a protective role played by the hormones testosterone and estrogen in preserving the structure and function of brain cells (Hart 2001). However, another study looking at histologic (i.e., microscopic, cellular) changes in the brain associated with aging in dogs did not support a link between neutering and cognitive decline (Waters et al. 2000). In her summary of potential impacts of neutering on the ability of working dogs, Margaret Root Kustritz (2014) reported that in general, neutering did not negatively impact working ability or trainability, and might even have positive impacts on trainability in some breeds (Serpell and Hsu 2005) and on safety of herding dogs due to a reduction in roaming behavior. In his book, *The Genius of Dogs*, researcher Brian Hare describes how neutering his seemingly-unruly chow chow "Milo" resulted in a much calmer, more focused, and more trainable dog (Hare and Woods 2013). Hare credits these marked improvements in temperament to the reduction in the levels of male reproductive hormones in Milo's body. More research is needed in this area.

Risks associated with the surgery itself

As with any surgery, there are risks of *complications* (adverse reactions to anesthesia, infection, etc.) associated with spaying and neutering your dog. These risks may be particularly worrisome for some owners who are unsure about whether or not to neuter their dogs, given that the vast majority of these surgeries are considered "elective" surgeries; that is, planned, non-emergency, optional procedures. As with any elective procedure, neutering should not be performed on any animal who is not healthy, or is not well able to tolerate anesthesia (Root Kustritz 2014). Advocates of "pediatric neutering" (neutering done between 6 and 16 weeks of age) note that puppies tolerate anesthesia and surgery well, have shorter surgery times than adult dogs (with shorter surgeries generally being safer for the animal involved), and recover quickly (Root Kustritz 2014).

A number of studies have looked at how commonly complications directly associated with the spay/neuter surgery occur in dogs. Reported incidence of postoperative complications was 6.1 percent in one survey of 1,016 dogs, with most of these considered minor problems such as inflammation at the surgery site and gastrointestinal tract upset (Pollari et al. 1996). Another study tabulated rates of complications during spay (ovariohysterectomy) surgery (6.3%), postoperative complications following the spay (14.1%), and total complications (20.6%) at one veterinary teaching hospital (Burrow et al. 2005). Serious complications resulting from neutering surgery include infections, abscesses, chewed-out sutures, or rupture of the surgical wound, and occur at an estimated frequency of 1–4 percent of cases (Pollari and Bonnett 1996). Spay surgeries are more invasive, and accounted for 90 percent of the serious complications in that study. The death rate associated with complications due to spay/neuter surgeries is relatively low (around 0.1%; Pollari et al. 1996).

In the US, where these surgeries are performed routinely, most veterinarians have extensive experience in performing the procedure. Routine, careful monitoring of patients under anesthesia is recommended by American College of Veterinary Anesthesiologists, and can help ensure the safety of your dog during surgery (Mazzaferro 2011). In any case, you can always ask your veterinarian how much experience he or she has in performing this surgery (or any other surgical procedure, for this matter).

The realities of living with an intact dog

Owners planning to keep their dogs intact should be familiar with the realities of living with an intact dog, in particular the intensity of reproductive drive in many of these dogs (both male and female), which can lead to increased escape attempts, increased attention from other neighborhood dogs, and in some cases, increased aggression. Owners should consider whether they can safely contain an intact dog who is determined to locate a mate. Many unplanned litters of puppies result when dogs manage to slip through "secure" fences and other boundaries.

Figure 9.3: The decision about whether or not to neuter your dog can be complex, but is less intimidating when viewed as a risks versus benefits analysis, discussed with your veterinarian, in light of your dog's breed and lifestyle. (Photo: ©Donna Kelliher, www.DonnaKelliher.com)

Box 9.5. To neuter or not to neuter?

What follows is a brief summary of the most common recommendations regarding spaying or neutering your dog. Given what is known (and not known) today about the medical risks and benefits of neutering, it is not really possible to make recommendations about what is best for each and every dog. Thus, the generalizations listed below are not meant to substitute for discussions with your veterinarian, conducted as part of a complete lifelong wellness plan for your dog, nor do they encompass every angle of the various risks and benefits of these surgeries, or every argument for and against neutering. But, hopefully they will serve as a compass to point you in the right direction: the direction that is best for you and your dog.

1. Dogs who do not have a permanent home (e.g., dogs in shelter care) should be neutered, prior to (or soon after) release from the shelter, given the number of homeless and unwanted dogs who are euthanized each year.

2. For dogs who do have a permanent home with a caring human family, the decision of whether or not to neuter should be made on an individual basis by the owner, in consultation with their dog's veterinarian; and in particular, taking into account factors like breed risks for certain types of cancers, breed/size of dog, and risk for joint problems such as hip dysplasia, and so on.

3. It may be advisable to delay neutering until the dog is physically mature, to avoid a number of health risks which seem to be associated with early neutering.

4. In general, health benefits of neutering are better supported for female dogs, versus male dogs.

To neuter or not to neuter: a summary

Are you confused yet about the pros and cons of neutering your canine companion? If so, this is understandable; there are many factors to weigh in making the decision, with the relative importance of these factors varying with breed, size of your dog, your lifestyle, and perhaps your own (or your family's) personal beliefs about this procedure. In addition, our understanding of the long-term impacts of neutering your dog is still evolving, as new research continues to be conducted and published on this issue. A number of authors have attempted to simplify and summarize the various risks and benefits of neutering your dog, and at the risk of oversimplifying these issues, Box 9.5 lists a brief overview of the most common recommendations regarding whether or not to neuter your pet. If you are still unsure, I would recommend reading the Sanborn (2007) paper, or the Root Kustritz (2007) paper, both of which are listed in the "References and additional resources" section of this chapter.

Choosing a veterinarian: building a relationship for the life of your dog

Selecting a veterinarian, particularly if you are new to the area or a first-time pet parent, can be a daunting task, not unlike choosing a doctor for your own medical needs. Organizations like the American Veterinary Medical Association, the British Veterinary Association, and the Humane

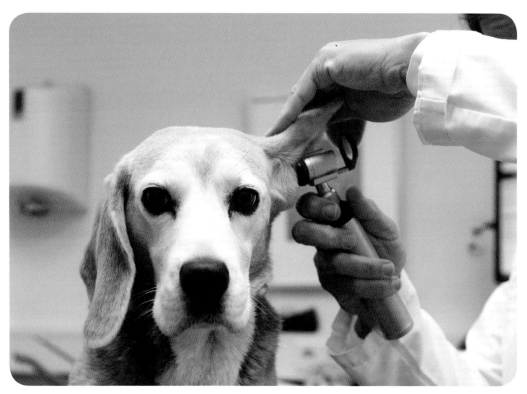

Figure 9.4: Choosing a veterinarian who works well with you and your dog is an important part of caring for your dog. (Photos: iStock)

Society of the United States have helpful pages with advice for locating a great veterinarian in your area.[12] Practical suggestions include:

- Talk to friends, neighbors, and other dog owners in your area, and get their recommendations for practices and veterinarians nearby that they really like. On the other hand, be wary of relying too heavily on consumer rating and social media websites like Yelp or Angie's List, as these tend to be more biased than they appear (Consumer Reports 2013); and, as is often the case with voluntary reporting services, tend to attract reviewers with the strongest negative experiences and opinions, or conversely, those with a vested interest in seeing that business succeed.
- Do not wait until you have a medical emergency with your dog to choose a veterinarian. Do your research ahead of time, and establish a relationship with the practice and veterinarian before a health crisis occurs. A good veterinary practice should allow you to bring your dog in, at a prearranged time, to meet the staff, get a few pats and treats, and become familiar with the location (without any of the stressors of an examination, vaccines, etc.), prior to your first real appointment.
- Choose a practice that is clean, well-maintained, with reception and waiting areas with sufficient space for you and your dog to wait without crowding other patients, and with courteous, friendly, and professional staff! If your dog has special needs (behavioral or medical), you may wish to inquire whether the practice has the specialty knowledge to handle these cases in-house, or to whom they would refer these issues.
- As mentioned earlier, optimal veterinary care occurs when you are able to work consistently with one primary care veterinarian (even if specialists are called in to deal with more serious or unusual medical issues). Practices vary in size from a single veterinarian to multi-doctor practices with 10 or more veterinarians; with larger practices, check to see if they support your requests to see the same veterinarian each time for regularly-scheduled appointments.
- Ask about on-call cover (i.e., if there is an emergency that happens outside of the clinic's regular business hours, who covers these? Is it one of the in-house veterinarians? Or are emergencies referred to an emergency clinic – and if so, where is that clinic located?). And finally, consider asking about the typical length of consultations, or appointment times; these can vary widely by clinic and veterinarian. Knowing how long the typical visit lasts may help you establish whether the veterinarian will be able to take what time is needed to fully discuss your dog's situation and symptoms, up to 30 minutes or more if necessary, or whether the clinic has a strict policy of short (5–15 minute) appointment windows.

Another aspect of veterinary practice that can vary widely is how the veterinarian and his or her staff interact with your pet (approach, restraint, handling, etc.). Low-stress handling and restraint

---

12 The AVMA's page on choosing a vet is at https://www.avma.org/public/YourVet/Pages/finding-a-veterinarian.aspx; the BVA's page is at http://www.bva.co.uk/You-and-your-vet/Choosing-a-vet/; and the Humane Society's page is at http://www.humanesociety.org/animals/resources/tips/choosing_a_veterinarian.html

Figure 9.5: Relaxed dog in the veterinary exam room. (Photo: P. Turmenne)

techniques (Yin 2009) are now the gold standard in veterinary medicine, and are designed to be sensitive to the emotional and behavioral needs of your dog (by limiting unnecessary fear and stress, by using the minimum amount of restraint necessary, by using treats and other positive rewards to reinforce calm behaviors and compliance with veterinary procedures), as well as to their physical needs. This movement is relatively new, however, and not all practitioners have adopted these techniques. As part of being an advocate for your dog's well-being, it is worth looking for a practice that advertises their use of these techniques (in the US, such practices often refer to themselves as "pet friendly practices"), or inquiring of a new veterinarian if he or she uses these techniques. This can be done either directly, by simply asking if and how these techniques are used in that practice; or indirectly, by asking what he or she usually does to help patients who are fearful during an exam. The next section discusses low-stress handling techniques in more detail.

Choosing a veterinarian: the importance of low-stress handling

At the start of this chapter, we recommended choosing a veterinarian who uses "low-stress handling" techniques (Yin 2009). This recommendation is also found in a remarkable number of recent veterinary guidelines and publications, such as the AAHA *Canine Life Stage Guidelines* (Bartges et al. 2012), the AAHA Canine and Feline Behavior Management Guidelines (Hammerle et al. 2015), the British Small Animal Veterinary Association's *BSAVA Manual of Canine and Feline Behavioural Medicine* (BSAVA 2012), Herron and Shreyer's 2014 *The Pet-Friendly Practice: A guide for practitioners*, and DVM360's special issue on "Facing Fear" and the "Fear Free"™ initiative spearheaded by Dr. Marty Becker (Lofflin 2014; Overall 2014).

So, what are "low-stress handling" techniques, and why are they so important for you and your dog? At their heart, these techniques are designed to build trust between the animal and the veterinarian, and to reduce the stress and fear associated with necessary veterinary care. Every veterinarian takes an oath upon graduation to prevent and relieve animal suffering, and low-stress handling is part of ensuring that veterinary care pays as much attention to the emotional well-being as it does to the physical well-being of non-human patients (Herron and Shreyer 2014). As the well-respected veterinarian and researcher Dr. Bonnie Beaver writes in her foreword to Dr. Yin's landmark book (Yin 2009):

> While every veterinarian ... will have slightly different approaches in how they positively handle the dog or cat, we all agree that it is time to get away from the 'brute-a-cane' approach of solving all problems. It does not work, the animal's fear escalates, and people and animals get hurt. Things get worse every time the animal returns to a similar situation. By taking a little extra time to encourage positive behavior and reduce fear, it is possible to relieve stress in the pets, the owners, and the veterinary staff.
>
> (Beaver 2009; p. 15)

For me, the best analogy to explain why these techniques make so much sense is one used by Dr. Yin herself – in a section of her book appropriately entitled, "A smarter alternative to force," she compares veterinary practices using low-stress handling techniques to pediatric dentists, who have built a niche for themselves based on understanding the behavior of their young human patients (Yin 2009; pp. 24–25). Why, Yin asks, do children misbehave at the dentist? Are they just being unruly and deliberately disruptive? In the vast majority of cases, no – they misbehave because they are fearful. Even if they were misbehaving for other reasons, would it be appropriate to strap them to the exam chair, or command them "in a military voice" to remain seated? Rather than trying to use force on their patients, pediatric dentists set out to provide these children with a fun experience at the clinic: they decorate the waiting and exam rooms with colorful pictures and murals, they provide toys, books, and video games for the children to play while they are waiting for their exam, they play familiar G-rated movies during exams, with screens set at an angle so that the patient can watch while the doctor works. Staff members are trained to know how to talk to and work with children, exuding seemingly endless patience and calmly explaining procedures before they begin. As Dr. Yin puts it, "They make a good first impression and maintain it throughout." (p. 24). My own 10-year-old son now goes to a dental practice very much like this one, and the results have been nothing short of remarkable: he has gone from a child very uncomfortable with even the thought of dental exams, to one who enjoys visits to the dentists and is motivated to take good care of his teeth. I have also, of course, seen similar results when accompanying fearful dogs to the veterinarian; with patience, proactive distraction, and calming techniques, and a healthy helping of delicious treats, it is amazing how well these dogs will do at the clinic.

For veterinarians dealing with canine patients, the most important aspects of low-stress handling are:

- The ability to accurately read canine body language.
- The willingness to listen to what these dogs are trying to tell us, and when necessary, to respond promptly and accordingly to reduce their fear.
- Familiarity with a range of restraint and handling tools and techniques that are less likely to dramatically increase fear or lead to defensive aggression. Examples of these include use of minimal restraint necessary; towel wraps instead of hug holds; "Calming Caps"[13] to limit

---

[13] The Calming- or ThunderCap is a mesh cover for the dog's head/eyes that limits, but does not prevent, vision; available from *http://www.thundershirt.com/thundercap.html*

visual stimulation; basket muzzles when necessary and with appropriate desensitization beforehand to wearing the muzzle; and distraction during procedures using high-value food treats or a favorite toy.
- Willingness to make adaptations as feasible to the reception/waiting area, the kennels/boarding area, and the exam room itself, in order to minimize environmental stressors and avoid the build-up of stress and fear that can occur if these aspects of the clinic are highly stressful places for dogs. Examples of such adaptations include ensuring sufficient space (and/or presence of visual barriers) in the waiting or reception area to avoid conflict between dogs, or between dogs and cats; putting a non-slip surface such as a bath mat or yoga mat onto the exam table to prevent the dog from slipping (or, doing the exam on the floor); stocking the exam room with a handy supply of high-value treats.
- Accurate record-keeping, to keep track of what handling/restraint/distraction approaches worked best for that dog in the past, of what procedures (if any) the dog has found particularly stressful or difficult to endure calmly, of any aggression shown and in what circumstances (to help staff track whether techniques used to decrease fear and aggression are working for this dog, and for the safety of all concerned).
- Strong client communication skills, taking whatever time is needed to collect a thorough medical history from the client, asking pertinent questions, and really listening to (and respecting) the client's concerns.

Ideally, your dog will see only low-stress, pet-friendly practitioners throughout his lifetime. Avoiding stressful veterinary experiences early in the puppy's life, and taking the time to build positive associations with the clinic location and staff, minimizes the risk of that dog developing a life-long fear of the veterinary clinic. All too commonly, fear experienced in the first puppy visits can grow and evolve over repeated visits, eventually resulting in a dog who is extremely stressed by veterinary visits, or who becomes defensively aggressive the moment that the veterinarian enters the exam room. These are situations that are stressful, and even dangerous, for all concerned – including the dog. As we discussed in the chapter on behavior and the importance of early socialization, it is much easier to prevent fear-related behavioral problems than to solve them once they are established. Seek out a veterinarian who is committed to building a positive, low-stress relationship with your dog, for the life of your dog.

> *Veterinarians must institute a culture of kindness in the practice and avoid using either forced restraint, or punitive training or management methods.*
>
> (Hammerle et al. 2015, p. 30)

"What if my dog is already afraid of the veterinarian"?

In some cases, we may find ourselves with a dog who is already afraid of the veterinarian, due to some past experience, and perhaps exacerbated by a shy or fearful nature. In these cases, what can we do to help our dog feel less stressed while receiving important veterinary care? What follows are some suggestions for reducing fear at the veterinary clinic, while maintaining safety of your dog and the veterinary staff.

- Begin to gradually acclimate your dog to the process of driving (or walking) to the veterinary office, entering the clinic, and interacting (in a friendly way) with the veterinary staff. Review the sections on systematic desensitization and counterconditioning in Chapter 7 of this book (these are the underlying learning principles you are applying here). Call your veterinarian's office and ask if and when it would be okay for you to bring your dog around, just to say hello (and perhaps get a few treats), without any work being performed on her.[14] If your dog is highly fearful (becoming fearful the moment she steps out of the car in the car park, for example), this may mean starting at a very low level of the trigger (in this case, a visit to the vet) and increasing the level of the trigger step by step, at your dog's own comfort level. For example, you might start by just driving to the clinic car park and having a few treats before leaving, and gradually working your way up to getting out of the car, approaching the clinic, entering the clinic reception area, entering the (empty) exam room, all the way to greeting the veterinarian in the exam room. If your dog only becomes stressed when the veterinarian first begins the physical exam, you can start much further along in the process, perhaps asking that the veterinarian simply offer a few pats and treats, and then heading home again. Repeat each step, as many times as your schedule (and the clinic's schedule) permits, until your dog is clearly comfortable with that level of the trigger, before increasing the level. In severe cases, a fast-acting anti-anxiety medication,[15] prescribed by your veterinarian and administered just prior to a veterinary visit, may be helpful in this process.
- If your dog is highly fearful, and/or if your dog has *ever* growled, lunged, or attempted to bite anyone at the veterinary hospital, begin immediately to countercondition and desensitize your dog to wearing a basket muzzle for veterinary visits. If done correctly, your dog will soon be very comfortable wearing a muzzle, even looking forward to it, if wearing a muzzle is accompanied by delicious treats inserted into the muzzle (like peanut butter, or spray cheese). Ensure that the muzzle you are using is sized and fitted correctly for your dog. In such cases, your dog should be wearing the muzzle before you even enter the veterinary clinic; not only will it prevent a bite, but knowing that your dog cannot bite will make staff calmer, which in turn may help your dog remain calmer. See Chapter 7 for more on desensitizing and counterconditioning your dog to a muzzle, including links to helpful videos illustrating the process. Many owners are upset by the idea of their beloved family dog wearing a muzzle, but it is important to remember that these are just safety tools, and are worn for the safety of all concerned, *including your dog*. And, it is *much* less stressful for your dog to gradually learn to tolerate the muzzle at home, working calmly with someone he trusts, than to be in a situation where a muzzle is forcibly applied at the veterinary clinic

---

14 Most clinics welcome these visits, provided that advance notice is given, as a stress-free chance to meet and greet your dog.

15 More information on these medications can be found in Chapter 7; note that acepromazine, commonly used for this purpose in the past, is no longer recommended for use in this way. More information on the use of acepromazine can also be found on Debbie Jacob's wonderful site "FearfulDogs.com": *http://fearfuldogs.com/acepromazine/*

after a growl or attempted bite. Remember, in the vast majority of cases, aggression seen at the veterinary clinic is motivated by fear.

- At the time of your appointment, be sure to go prepared, with your dog on a secure lead and collar or head halter, and armed with a healthy supply of treats preferred by your dog (unless your dog is scheduled for surgery on that day, in which case you will have to forego feeding any food prior to the surgery). The key here is to begin distracting your dog *before* he has a chance to become stressed! For many dogs, this means starting the work the moment you enter the clinic. The most common mistake I see in this situation is an owner waiting until their dog is already fearful (i.e., showing clear body language signs of fear) before beginning to try and distract him with cues and treats. By this point, it is usually too late to intervene, and distraction will not be nearly as effective, if it works at all. Distraction can be as easy as asking your dog to focus on you, "look at that," sit, or lie down (and, rewarding your dog each time he complies with a delicious treat). If your wait for your appointment is longer than a few minutes, your dog may benefit from walking on-leash with you around outside the clinic while waiting, if the area immediately surrounding the clinic is safe for dog-walking. Ask the veterinary staff to offer your dog a few treats (tossing them from a few feet away if necessary) prior to approaching or touching your dog; this can be done while you are giving the veterinarian or technician your dog's medical history, and so on. Throughout this process, it is important to keep an eye on your dog's body language; if you sense that your dog is becoming stressed, let the veterinarian know and ask if it would be possible to take a break for a minute or two to allow your dog to regain his calm. If at any point during the procedure your dog suddenly stills or becomes very tense, growls, or bares his teeth, immediately and calmly remove your dog to a safe distance. It may then be necessary to reschedule the rest of the appointment for another day (and consider using a muzzle, as directed above).

Alternative veterinary medicine: what's the scoop?

Just as with human medicine, many canine caretakers are interested in investigating the use of alternative therapies in their dog's medical care. Alternative therapies include approaches such as acupuncture, aromatherapy, chiropractic, and medicinal herbs, and practitioners using these therapies may integrate them with other, modern-medicine drugs and surgical techniques. This is a somewhat controversial area, with opinions about the effectiveness of alternative therapies varying widely within the veterinary community. The AVMA's position states that (among other things) "all aspects of veterinary medicine should be held to the same standards, including complementary, alternative and integrative veterinary medicine, non-traditional or other novel approaches" and "Diagnosis and treatment should be based on sound, accepted principles of veterinary medicine and the medical judgement of the veterinarian … and, veterinarians should have the requisite knowledge and skills for every treatment modality they consider using." The American Holistic Veterinary Medical Association has a website (www.ahvma.org) and a peer-reviewed journal providing more information on these therapies. The AHVMA website also provides a search tool to help owners locate a holistic or integrative medicine veterinarian in their area. In Britain, the British Association of Homeopathic Veterinary Surgeons (www.bahvs.com)

provides many of the same resources. I would encourage you, if you are interested in learning more about these therapies, to talk to your veterinarian, and do your own research into the safety and efficacy of these treatments. Anecdotally, I have had dog owners and caretakers tell me how these therapies have helped their pets, so they are worth investigating.

Recap: veterinary care for the life of your dog

To recap, the emphasis in veterinary medicine today is on preventative and wellness care, individualized care, and building a strong partnership with your dog's primary care veterinarian in providing for the medical needs of your canine companion. This chapter has presented an overview of a broad, complex, and often confusing (to non-veterinarians, at least) topic, but we hope that this overview has been helpful in giving you a sense for the state of the science. We encourage you to consult the recommended resources and references for more in-depth information, should you so desire. And, of course, always consult with your veterinarian about any health questions or concerns arising with your canine family members.

Take-home messages

1. An important part of bringing a dog into your home is assuming the responsibility for providing that dog with all necessary veterinary care, for the life of that dog. The potential costs associated with this care should be considered before adopting a dog.
2. Choose a skilled, professional, and compassionate primary care veterinarian for your dog, and if possible, work with this same veterinarian for the life of your dog. Ideally, this veterinarian will work at a practice that follows and endorses low-stress handling and restraint techniques, to reduce any fear that your dog may feel associated with veterinary visits and procedures.
3. Core vaccines are recommended for all puppies, and dogs of unknown vaccine history. Non-core vaccines may or may not be necessary for your dog, depending on where you live and other factors. Discuss vaccines and vaccine schedules with your veterinarian, with the goal being to minimize known vaccine risks while ensuring that your dog gets the proven benefits of vaccination.
4. Neutering dogs has both demonstrated benefits and risks. The decision to neuter owned companion dogs should be made by the dog's owner/caretaker, in consultation with the veterinarian, and taking into account the most current information on the risks and benefits of neutering. In particular, if your dog is a purebred, ask your veterinarian or breeder about breed predilections for certain diseases, as risks and benefits of neutering can vary markedly with breed.

References and additional resources

Arnold, S., Arnold, P., Hubler, M., Casal, M., & Rüsch, P. (1989). [Urinary incontinence in spayed female dogs: Frequency and breed disposition]. *Schweizer Archiv fur Tierheilkunde* 131: 259–263.

Bartges, J., Boynton, B., Hoyumpa Vogt, A., Krauter, E., Lambrecht, K., Svec, R., & Thompson, S. (2012). AAHA canine life stage guidelines. *Journal of the American Animal Hospital Association* 48: 1–11.

Beauvais, W., Cardwell, J.M., & Brodbelt, D.C. (2012a). The effect of neutering on the risk of mammary tumours in dogs – a systematic review. *Journal of Small Animal Practice* 53: 314–322.

Beauvais, W., Cardwell, J.M., & Brodbelt, D.C. (2012b). The effect of neutering on the risk of urinary incontinence in bitches – A systematic review. *Journal of Small Animal Practice* 53: 198–204.

Beaver, B. (2009). Foreword. In Yin, S., *Low Stress Handling, Restraint and Behavior Modification of Dogs & Cats: Techniques for developing patients who love their visits*. Davis, CA: Cattledog Publishing. p 15.

British Small Animal Veterinary Association (BSAVA) (2012). BSAVA Manual of Canine and Feline Behavioural Medicine (2nd ed.). D.F. Horwitz and D.S. Mills (Eds.). Gloucester, UK: The British Small Animal Veterinary Association. 324 p.

Burrow, R., Batchelor, D., & Cripps, P. (2005). Complications observed during and after ovariohysterectomy of 142 bitches at a veterinary teaching hospital. *The Veterinary Record* 157: 829–833.

Consumer Reports (2013). The truth about online ratings services. *Consumer Reports Money Advisor*, October 2013. Available at http://www.consumerreports.org/cro/index.htm

Cooley, D.M., Beranek, B.C., Schlitter, D.L., Glickman, L.T., & Waters, D.J. (2002). Endogenous gonadal hormone exposure and bone sarcoma risk. *Cancer Epidemiology, Biomarkers & Prevention* 11: 1434–1440.

Davis-Wurzler, G.M. (2014). 2013 update on current vaccination strategies in puppies and kittens. *Veterinary Clinics of North America: Small Animal Practice* 44: 235–263.

de Bleser, B., Brodbelt, D.C., Gregory, N.G., & Martinez, T.A. (2011). The association between acquired urinary sphincter mechanism incompetence in bitches and early spaying: A case-control study. *Veterinary Journal* 187: 42–47.

Dobson, J.M., Smauel, S., Milstein, H., et al. (2002). Canine neoplasia in the U.K.: Estimates of incidence rates from a population of insured dogs. *Journal of Small Animal Practice* 43: 240–246.

Dorn, C.R., Taylor, D.O.N., Schneider, R., et al. (1968). Survey of animal neoplasms in Alameda and Contra Costa counties, California. II. Cancer morbidity in dogs and cats from Alameda county. *Journal of the National Cancer Institute* 40: 307–318.

Duval, J.M., Budsberg, S.C., Flo, G.L., et al. (1999). Breed, sex, and body weight as risk factors for rupture of the cranial cruciate ligament in young dogs. *Journal of the American Veterinary Medical Association* 215: 811–814.

Edney, A.T., & Smith, P.M. (1986). Study of obesity in dogs visiting veterinary practices in the United Kingdom. *The Veterinary Record* 118: 391–396.

Egenvall, A., Bonnett, B.N., Ohagen, P., et al. (2005). Incidence and survival after mammary tumors in a population of over 80,000 insured female dogs in Sweden from 1995–2002. *Journal of Small Animal Practice* 69: 109–127.

Forsee, K.M., Davis, G.J., Mouat, E.E. et al. (2013). Evaluation of the prevalence of urinary incontinence in spayed female dogs: 566 cases (2003–2008). *Journal of the American Veterinary Medical Association* 242: 959–962.

Hagman, R., Lagerstedt, A.S., Hedhammer, A., et al. (2011). A breed-matched case-controlled study of potential risk factors for canine pyometra. *Theriogenology* 75: 1251–1257.

Hammerle, M., Horst, C., Levine, E., Overall, K., et al. (2015). AAHA canine and feline behavior management guidelines. *Journal of the American Animal Hospital Association* 51: 205–221.

Hare, B., & Woods, V. (2013). *The Genius of Dogs*. New York: Plume/Penguin Group USA. 367 p.

Hart, B.L. (2001). Effect of gonadectomy on subsequent development of age-related cognitive impairment in dogs. *Journal of the American Veterinary Medical Association* 219: 51–56.

Hart, B.L., Hart, L.A., Thigpen, A.P., Willits, N.H. (2014). Long-term health effects of neutering dogs: Comparison of Labrador retrievers with golden retrievers. *PLOS ONE* 9: e102241 1–10

Herron, M. E., & Shreyer T. (2014). The pet-friendly veterinary practice: A guide for practitioners. *Veterinary Clinics of North America: Small Animal Practice* 44: 451–481.

Hunthausen, W.K. (2010). Preventative behavioural medicine for dogs. In D. Horwitz, D. Mills, and S. Heath (Eds.). *BSAVA Manual of Canine and Feline Behavioural Medicine* (2nd ed.) Gloucester, UK: British Small Animal Veterinary Association. pp. 65–66.

Johnston, S.D., Root Kustritz, M.V., & Olson, P.N. (2001). Disorders of the canine uterus and uterine tubes (oviducts). In S.D. Johnston, M.V. Root Kustritz, and P.N. Olson (Eds.). *Canine and Feline Theriogenology*. Philadelphia, PA: WB Saunders. pp. 206–224.

Lofflin, J. (2014). Facing Fear. *dvm360* 45: 8–22.

Mazzaferro, E. (2011). Anesthesia monitoring: Raising the standards of care. *Clinician's Brief*, August 2011: 51–54.

McGreevy, P.D., Thomson, P.C., Pride, C., Fawcett, A., Grassi, T., & Jones, B. (2005). Prevalence of obesity in dogs examined by Australian veterinary practices and the risk factors involved. *The Veterinary Record* 156: 695–702.

Moe, L. (2001). Population-based incidence of mammary tumours in some dog breeds. *Journal of Reproduction and Fertility Supplement* 57: 439–443.

National Council of Pet Population Study and Policy (1994). *National Shelter Census: 1994 results*. Fort Collins, CO: National Council on Pet Population Study and Policy.

Nibblett, B.M., Ketzis, J.K., & Grigg, E.K. (2014). Comparison of stress exhibited by cats examined in a clinic versus a home setting. *Applied Animal Behaviour Science* 173: 68–75.

Niskanen, M., & Thrusfield, M.V. (1998). Associations between age, parity, hormonal therapy and breed, and pyometra in Finnish dogs. *The Veterinary Record* 18: 493–498.

Overall, K. (2014). Your complete guide to reducing fear in veterinary patients. *dvm360* 45: 24–36.

Pollari, F.L., & Bonnett, B.N. (1996). Evaluation of postoperative complications following elective surgeries of dogs and cats at private practices using computer records. *Canadian Veterinary Journal* 37: 672–678.

Pollari, F.L., Bonnett, B.N., Bamsey, S.C., et al. (1996). Postoperative complications of elective surgeries in dogs and cats determined by examining electronic and paper medical records. *Journal of the American Veterinary Medical Association* 208: 1882–1886.

Priester, W.A., & McKay, F.W. (1980). The occurrence of tumors in domestic animals. National Cancer Institute Monograph 54: 269.

Prymak, C., McKee, L.J., Goldschmidt, M.H., et al. (1985) Epidemiologic, clinical, pathologic and prognostic characteristics of splenic hemangiosarcoma and splenic hematoma in dogs: 217 cases. *Journal of the American Veterinary Medical Association* 193: 706–712.

Reichler, I.M. (2009). Gonadectomy in dogs and cats: A review of risks and benefits. *Reproduction in Domestic Animals* 44 (Suppl. 2): 29–35.

Root Kustritz, M. (2007). Determining the optimal age for gonadectomy of dogs and cats. *Journal of the American Veterinary Medical Association* 231: 1665–1675.

Root Kustritz, M. (2010). Optimal age for gonadectomy in dogs and cats. *Clinical Therapeutics* 2: 177–181.

Root Kustritz, M. (2014). Pros, cons and techniques of pediatric neutering. *Veterinary Clinics of North America: Small Animal Practice* 44: 221–233.

Ru, G., Terracini, B., & Glickman, T. (1998). Host-related risk factors for canine osteosarcoma. *The Veterinary Journal* 156: 31–39.

Salmeri, K.R., Bloomberg, M.S., Scruggs, S.L., et al. (1991). Gonadectomy in immature dogs: Effects on skeletal, physical and behavioral development. *Journal of the American Veterinary Medical Association* 198: 1193–1203.

Sanborn, L.J. (2007.) Long-term health risks and benefits associated with spay/neuter in dogs. New Brunswick, NJ: Rutgers University, Dept. of Animal Sciences. (unpublished manuscript). Available at *www.naiaonline.org/pdfs/longtermhealtheffectsofspayneuterindogs.pdf*

Schneider, R., Dorn, C.R., Taylor, D.O. (1969). Factors influencing canine mammary cancer development and post-surgical survival. *Journal of the National Cancer Institute* 43: 1249–1261.

Serpell, J.A., & Hsu, Y. (2005). Effects of breed, sex and neuter status on trainability in dogs. *Anthrozoos* 18: 196–207.

Shiel, R.E., Puggioni, A., & Keeley, B.J. (2008). Canine urinary incontinence. Part 2. Treatment. *Irish Veterinary Journal* 61: 835–840.

Smith, A.N. (2014). The role of neutering in cancer development. *Veterinary Clinics of North America: Small Animal Medicine* 44: 965–975.

Sorenmo, K.U., Goldschmidt, M., Shofer, F., & Ferrocone, J. (2003). Immunohistochemical characterization of canine prostatic carcinoma and correlation with castration status and castration time. *Veterinary Comparative Oncology* 1: 48–56.

Spain, C.V, Scarlett, J.M., & Houpt, K.A. (2004). Long-term risks and benefits of early-age gonadectomy in dogs. *Journal of the American Veterinary Medical Association* 224: 380–387.

Stepita, M., Bain, M.J., & Kass, P.H. (2013). Frequency of CPV infection in vaccinated puppies that attended puppy socialization classes. *Journal of the American Animal Hospital Association* 49: 95–100.

Stöcklin-Gautschi, N.M., Hassig, M., Reichler, I.M., Hubler, M., & Arnold, S. (2001). The relationship of urinary incontinence to early spaying in bitches. *Journal of Reproduction and Fertility (Supplement)* 57: 233–236.

Teske, E., Naan, E.C., van Dijk, E.M., van Garderen, E., & Schalken, J.A. (2002). Canine prostate carcinoma: Epidemiological evidence of an increased risk in castrated dogs. *Molecular and Cellular Endocrinology* 197: 251–255.

Thrusfield, M.V., Holt, P.E., & Muirhead, R.H. (1998). Acquired urinary incontinence in bitches: Its incidence and relationship to neutering practices. *Journal of Small Animal Practice* 39: 559–566.

Torres de la Riva, G., Hart, B.L., Farver, T.B., Oberbauer, A.M., et al. (2013). Neutering dogs: Effects on joint disorders and cancers in golden retrievers. *PLOS ONE* 8: e55937.

University of California at Davis (UC Davis) (2012) *VMTH Canine and Feline Vaccination Guidelines*. http://www.vetmed.ucdavis.edu/vmth/small_animal/internal_medicine/newsletters/vaccination_protocols.cfm (accessed May 2016).

Volk, J.O., Felstad, K.E., Thomas, J.G., & Siren, C.W. (2011). Executive summary of the Bayer veterinary care usage study. *Journal of the American Veterinary Medical Association* 238: 1275–1282.

Ware, W.A., & Hopper, D.L. (1999). Cardiac tumors in dogs: 1982–1995. *Journal of Veterinary Internal Medicine* 13: 95–103.

Waters, D.F., Shen, S., & Glickman, L.T. (2000). Life expectancy, antagonistic pleiotropy, and the testis of dogs and men. *Prostate* 43: 272–277.

Weaver, A.D. (1981). Fifteen cases of prostatic carcinoma in the dog. *The Veterinary Record* 190: 71–75.

Wellborn, L.V., DeVries, J.G., Ford, R., Franklin, R.T., Hurley, K.F., McClure, K.D., Paul, M.A., & Schultz, R.D. (2011). AAHA canine vaccination guidelines. *Journal of the American Animal Hospital Association* 47: 1–42.

White, C.R., Hohenhaus, A.E., Kelsey, J., & Procter-Gray, E. (2011).Cutaneous MCTs: Associations with spay/neuter status, breed, body size, and phylogenetic cluster. *Journal of the American Animal Hospital Association (AAHA)* 47: 210–216.

Whitehair, J.G., Vasseur, P.B., & Willits, N.H. (1993). Epidemiology of cranial cruciate ligament rupture in dogs. *Journal of the American Veterinary Medical Association* 203: 1016–1019.

Yin, S. (2009). *Low Stress Handling, Restraint and Behavior Modification of Dogs & Cats: Techniques for developing patients who love their visits*. Davis, CA: Cattledog Publishing. 469 p.

Zink, M.C., Farhoody, P., Elser, S.E., Ruffini, L.D., Gibbons, T.A., & Rieger, R.H. (2014). Evaluation of the risk and age of onset of cancer and behavioral disorders in gonadectomized Vizslas. *Journal of the American Veterinary Medical Association* 244(3): 309–319. doi: 10.2460/javma.244.3.309.

Chapter 10

Too short lives: quality of life and end-of-life decisions for your dog

Photo: iStock

MAKING QUALITY OF LIFE DECISIONS AND END-OF-LIFE DECISIONS IS THE TOUGHest part of being a dog caregiver. It is our role to ensure that our dogs' needs are recognized and respected. Quality of life is a way to see how or if a dog's changing needs are being met and to determine whether or not our efforts to preserve his life are enhancing or are hindering the life of a happy dog. I vividly remember the day I realized that my sweet rescue dog Mojo, was no longer experiencing a life of happiness. I had not been prepared. She was only 13, had been a healthy and vibrant German shepherd dog mix, and had been my closest furry companion since I had fostered her when she was just 10 months old. Mojo was a behavior project; deemed unadoptable, left untrained, and being evaluated for euthanasia at my local

Figure 10.1: Senior dogs. (Photos: Adobe Stock)

shelter. I was called in to evaluate her suitability for adoption, given that she had mouthed a volunteer so aggressively that she bruised the woman's arm from wrist to shoulder through a heavy winter coat. She was a crazed dog inside of the floor-to-ceiling chain link-walled, cement-floored kennel in the tiny, crowded shelter building. She had never known the comfort of a home or the kindness of a loving caretaker. Mojo chewed at the kennel wall out of frustration and anxiety then jumped repeatedly on the kennel door, barking madly. When I leashed her she jumped on me incessantly, mouthing me hard all the while; when I turned my back she mouthed the back of my head. Clearly, it was not going well for Mojo or for me, the newly minted graduate student of animal behavior. Almost out of desperation, I led Mojo into the office for a bit of respite. It was there that Mojo's personality was truly set free. She stopped moving, stopped pulling, stopped jumping, and just laid her head on my lap and looked up at me almost pleadingly. There was a true companion dog in there! I took Mojo home, fostered her, and later adopted her. It took a few years and countless hours of training, but Mojo became a well-behaved, loving, and loyal companion. I somehow felt that she would continue along by my side forever. That is why, even though I saw the signs of canine cognitive dysfunction, I dismissed them. I thought maybe she was just nervous about our recent move or maybe she was panting from the heat. My veterinarian found nothing physically wrong with her but her cognitive decline became more evident each day. Mojo would wander around the house aimlessly, looking lost while standing in a very familiar hallway panting. It was still a great shock to me the day that my husband called me, panicked; Mojo had fallen in the laundry room and simply could not or would not get up. We rushed her to the emergency clinic where they ran every test possible, only to find out there was nothing physically (at least as they could tell) causing her symptoms. But, I knew I was losing her. Mojo stopped eating and rarely drank. But that day, the day I will never forget, she did not want to get up and go outside and she barely made eye contact with me; she whined and panted and panted. I knew it was time to say goodbye and to let her go. I cannot help but think it would have been easier had I been more prepared, more aware, or more … anything. The fact is, it is never an easy decision to euthanize your dog and it is no easier to lose your dog suddenly or to natural causes. There are some tools that can help you assess when it is the right time for you and your dog to say goodbye, and there are resources to help you cope during this most difficult time. In this chapter we will look to tools such as scales to assess the quality of life of your ill or aging dog, resources for coping with pet loss and grief, options for euthanasia, including the most difficult choice: euthanasia for severe behavior issues. Then we will explore grief support and how to move on after the loss of your dog.

Knowing when it is time to say goodbye

The veterinarian helping Mojo pass, who specialized in in-home euthanasia,[1] told me that no one ever feels it is the exact right time to let go of his or her dog. Caregivers feel it is either too early or too late, but never the exact right time. This resonated with me. For Mojo it at first seemed too early, as she had a few good moments. But then, waiting for the scheduled euthanasia day to arrive, as I watched her deteriorate, I began to think that I had waited too long. Deciding the time to compassionately end the life of your dog (euthanasia) is one of the most difficult decisions you will ever make. Part of providing an enriched, happy life is knowing when and how to end suffering.

It is normal to feel guilty and regretful, and to worry that you are not making the best decision for your dog, but take it easy on yourself. Chances are that you have provided your dog with the best life possible, ensuring he had the Five Freedoms (see Box 10.1 to see if your dog is still enjoying these Freedoms), and it was your caring concern that brought you to this day. Determining when to let your dog pass is emotionally challenging but getting an objective perspective can aid

---

1 See Box 10.7 for directories of veterinarians that perform in-home euthanasia.

> ## Box 10.1. Considering the Five Freedoms* and quality of life determinations
>
> *Freedom from hunger and thirst*: Does your dog have the appetite and motivation to eat and drink enough to maintain good health and vigor?
>
> *Freedom from discomfort*: Are you able to provide for your dog so that he is free from discomfort? With an aging/ill dog, are you able to adjust your environment so that your dog has a comfortable place to live? If you have stairs and he has limited mobility, can you accommodate him? If your dog is a risk to children or the elderly, can you assure his safety and theirs? If he needs to be managed for safety, does he have the space to seek comfort?
>
> *Freedom from pain, injury or disease*: Can you safely and effectively manage his pain? Can you afford the costs associated with the treatment and management of his injury or disease? Is his disease or injury manageable sufficiently to avoid suffering?
>
> *Freedom to express normal behavior*: Is your dog able to exercise, play, explore, and be social? If your dog is aggressive, can he be managed so that he has adequate space to practice natural behaviors?
>
> *Freedom from fear and distress*: Is your dog able to cope with changes associated with aging and disease? Does he find comfort in the things that once soothed his anxiety or is he constantly stressed?
>
> *Source: UK Farm Animal Welfare Council, 1965 and 2009.

in decision-making. There are tools to help caregivers navigate through this decision; one tool that helps determine what is best for your dog is a quality of life assessment.

Determining quality of life

As discussed in Chapter 2, quality of life (QoL) can be defined in multiple ways. It can be as simple as the general well-being of your dog or the health, comfort, and happiness experienced by your dog. It could also be characterized by your dog's ability to perform their natural or normal everyday behaviors. QoL can also be measured in different ways. One way of defining QoL is through observations by caregivers and veterinary professionals' recognizing and interpreting behavior or other health parameters, to determine if the dog is within normal limits. This form of assessment can give you a general idea of the QoL of your dog, but it can be subjective, because it is dependent on the caretaker's or veterinary professional's observations and reporting of the dog's behavior (see Box 10.2). Other QoL assessments are based on quantitative measurements of observable behavior such as signs of chronic and physiological stress. Questions on these assessments vary but they are generally scaled questions about the dog's level of pain and ability to perform everyday functions (Table 10.1). Some of these assessments have been tested for validity such as the Canine Brief Pain Inventory, which is a 10-item scaled questionnaire (Brown et al. 2007; 2009). Assessments such as this one have been shown to respond to changes in the health of the dog as a disease progresses. These assessments also can detect improvements in pain in response to treatment (Brown et al. 2008). Research on health-related QoL assessments has shown that these tools are purposeful and useful to aid treatment and in decision making (Freeman et al. 2005; Lynch et al. 2011; Reid et al. 2013).[2] Another benefit to QoL assessment is that it can improve communication and continuity of care between you and your veterinary professional. QoL evaluation can help clearly define where your dog is in his treatment and help determine the prognosis. As medical treatments advance, caregivers have greater need for QoL assessment to elect what course of treatment is the best for their furry companion (see Table 10.1).

Canine decline and cognitive dysfunction

> **Box 10.2. Quality of life determination through observation**
>
> - Remember how your pet looked and behaved prior to the illness. Sometimes changes are gradual and therefore hard to recognize. Look at photos or videos of your pet from before the illness.
> - Mark good and bad days on a calendar. This could be as simple as a happy or sad face for good or bad. If the bad days start to outweigh the good, it may be time to discuss euthanasia.
> - Write a concrete list of three to five things your pet likes to do. When your pet is no longer able to enjoy these things, it may be time to discuss euthanasia.
>
> Source: Nielson, J. (2013). How do I know it is time? The Ohio State University Veterinary Medical Center.

As health care continues to improve, dog caregivers are faced with complex issues related to advanced age. One condition that is more prevalent in recent years (or at least more recognized) is cognitive dysfunction syndrome (CDS). CDS is a degenerative condition that presents as a group of symptoms related to the aging of the canine brain (Box 10.3). These brain changes lead to cognitive impairments in memory and learning as well as social changes, disorientation, lapses of normal house-training, and changes in sleep and activity. These symptoms are described by the acronym DISH or DISHA (Table 10.2 and Table 10.3). In addition, dogs with CDS commonly display signs of fear, phobias, and anxiety (Table 10.3).

The greatest risk factor for CDS is age: 28 percent of dogs aged 11–12 years show at least one sign of CDS and by 16 years of age, 68 percent of dogs show at least one sign. Although CDS is becoming more widely recognized, data on the prevalence of CDS suggest that the phenomenon is underestimated in veterinary medicine (Osella et al. 2007). Be watchful for signs of CDS in your aging dog and tell your veterinarian if you suspect your dog might be experiencing CDS. Symptoms of CDS gradually worsen over time (Madari et al. 2015; Landsberg et al. 2013; Osella et al. 2007). The disease is progressive and dogs with at least one sign will show additional signs over the next 12 months. Although there is currently no cure for CDS, early detection and treatment is the best method for slowing the degenerative process, and maximizing QoL for the remainder of the dog's life.

Treatment for CDS is aimed at slowing the advancement of neuronal damage and cell death

---

2 A thorough review of published QoL assessments revealed that the current scales that have been compiled are suitable for dogs with a single disease. Many dogs have multiple conditions so further validation studies are needed to see if the QoL assessment is appropriate in all cases (Belshaw et al. 2015).

Table 10.1 How do I know when it's time? Assessing quality of life for your companion animal and making end-of-life decisions.

My pet …	Poor quality of life ←—————————→ Good quality of life				
	Strongly agree (all the time) (severe)	Agree (most of the time) (significant)	Neutral (sometimes) (mild)	Disagree (occasionally) (slight)	Strongly disagree (never) (none)
does not want to play	1	2	3	4	5
does not respond to my presence or does not interact with me in the same way as before	1	2	3	4	5
does not enjoy the same activities as before	1	2	3	4	5
is hiding	1	2	3	4	5
demeanor/behavior is not the same as it was prior to diagnosis/illness	1	2	3	4	5
does not seem to enjoy life	1	2	3	4	5
has more bad days than good days	1	2	3	4	5
is sleeping more than usual	1	2	3	4	5
seems dull and depressed	1	2	3	4	5
seems to be or is experiencing pain	1	2	3	4	5
is panting (even while resting)	1	2	3	4	5
is trembling or shaking	1	2	3	4	5
is vomiting and/or seems nauseous	1	2	3	4	5
is not eating well (may only be eating treats or only if fed by hand)	1	2	3	4	5
is not drinking well	1	2	3	4	5
is losing weight	1	2	3	4	5

is having diarrhea often	1	2	3	4	5
is not urinating well	1	2	3	4	5
is not moving normally	1	2	3	4	5
is not as active as normal	1	2	3	4	5
does not move around as needed	1	2	3	4	5
needs my help to move around normally	1	2	3	4	5
is unable to keep self clean after soiling	1	2	3	4	5
has coat that is greasy, matted, or rough-looking	1	2	3	4	5
How is my pet's overall health compared to the initial diagnosis/illness?	1 Worse	2	3 Same	4	5 Better

Current quality of life (place "X" along the line that best fits your pet's quality of life)

Poor	Good

Higher numbers on this chart equal a better quality of life. In some cases, even one item on the left-hand side of the chart (for example: pain) may indicate a poor quality of life, even if many of the other items are still positive. Some items or symptoms on the list may be expected side effects of the treatments that your pet is undergoing. It is important to discuss these symptoms and side effects with your veterinarian.

* Nielson, J. (2013). *How Do I Know it is Time?* The Ohio State University Veterinary Medical Center.

Table 10.2 Symptoms of cognitive dysfunction syndrome.

D	I	S	H
Disorientation/ confusion	*Interactions*	*Sleep–wake cycles*	*House soiling*
• Gets stuck or cannot get around objects • Stares blankly at walls or floor • Decreased recognition of familiar people/pets • Goes to wrong side of door; walks into door/walls • Drops food/cannot find food • Decreased response to auditory or visual stimuli • Increased reactivity to auditory or visual stimuli (barking)	• Decreased interest in petting/avoids contact • Decreased greeting behavior • In need of constant contact, "clingy" • Altered relationships with pets — less social/more irritable/aggressive • Altered relationships with people — less social/more irritable/aggressive	• Restless sleep/waking at nights • Increased daytime sleep	• Indoor elimination at sites previously trained • Decrease/loss of signaling • Goes outdoors, then returns indoors and eliminates • Elimination in crate or sleeping area

Adapted from Landsberg et al. (2012).

Note: Signs of CDS are summarized by the acronym DISH, which refers to disorientation; alterations in interactions with owners, other pets, and the environment; sleep–wake cycle disturbances; house soiling. Some authors add an "A" to the acronym (DISHA) for Activity level changes (i.e., initial reduced activity, eventually transitioning to increased, purposeless, or restless activity); see also Table 10.3.

Box 10.3. What causes cognitive dysfunction syndrome?

- Multiple neurological changes
- Unclear which specific changes cause which clinical signs
- Decline in neurons, changes in circulation
- Neurotoxic deposits
- Correlations between amount of beta-amyloid (neurotoxic protein) in the cerebral cortex and decline in cognitive ability
- Toxic free radicals (reactive oxygen species) increase with age as a result of stress, illness
- Correlated with cognitive decline
- Complicated by decreased cerebral vascular blood flow
- Toxins accumulate, blood flow reduced, neurons degenerate so neurotransmission compromised

Adapted from Landsberg, G., Hunthausen, W., & Ackerman, L. (2013). Handbook of Behavior Problems of the Dog & Cat (3rd ed.). New York: Elsevier, Ltd. 454 p.

Table 10.3 Changes in activity, anxiety, and learning resulting from cognitive dysfunction syndrome (CDS).

Activity: increased/ repetitive	*Activity: apathy/ depressed*	*Anxiety*	*Learning and memory*
• Pacing/wanders aimlessly	• Decreased interest in food/treats	• Vocalizations, restlessness/ agitation	• Decreased ability to perform learned tasks, commands
• Snaps at air/licks air	• Decreased exploration/activity/ play	• Anxiety, fear/phobia to auditory or visual stimuli	• Decreased responsiveness to familiar commands and tricks
• Licking owners/ objects	• Decreased self-care (hygiene)	• Anxiety, fear/phobia of places (surfaces, locations)	• Inability/slow to learn new tasks
• Increased appetite (eats quicker or more food)		• Anxiety/fear of people Separation anxiety	

Adapted from Landsberg et al. (2012) Signs of CDS in activity, anxiety, and learning.

Box 10.4. Management and treatment of cognitive dysfunction syndrome

- Increase physical and mental activity
- Incorporate enrichment: scent walks, play time, human interaction, puzzle toys
- Reinforce calm, quiet behaviors
- Ignore anxiety-based pacing and panting (avoid reinforcing)
- Increase access and opportunities for appropriate elimination: more trips outside
- If increased irritability leads to aggression, *avoid triggers*
- If vigilance increases, do not let dog wander too far, dog may get lost easily
- Diet: Hill's Prescription Diet b/d™ has been shown to improve memory, learning ability, clinical signs of CDS
- Ask your veterinarian about medications and nutritional supplements that can help with the symptoms of CDS

Source: Dodd et al. 2003. Can a fortified food affect behavioral manifestations of age-related cognitive decline in dogs? Veterinary Medicine 98: 396–408.

and improving clinical signs (see Table 10.2 and Table 10.3). Exercise, and novel toys can help to maintain cognitive function (Landsberg et al. 2012; see Box 10.4 for management and treatment of symptoms). Environmental enrichment including toys, exercise, and environmental stimuli, can have positive effects on behavioral health and quality of life in pets and is likely to improve cognitive function (Dodd et al. 2003). This is analogous to human studies in which increased mental activity and physical exercise have been found to delay the onset of dementia. Enrichment should focus on positive social interactions and stimulating ways to obtain food and treats. Food toys that require pushing, lifting, dropping, batting, pawing, or rolling to release food help older dogs to remain active and alert (see Chapter 6 for more ideas on environmental enrichment).

The following should be considered in pets that show signs of CDS or cognitive age-related behaviors in order to provide the happiest life possible. Inconsistency in the dog's environment can cause stress and negatively impact health and well-being (Landsberg et al. 2013). Avoid changes in the environment as much as possible. As sensory, motor, and cognitive function decline, new odor, tactile, and/or sound cues may help the pet better cope with its environment. Consider mats with different textures for eating places or sleeping places. Other cues may include bells on doors and different odors for different rooms. Dogs may need more frequent trips outdoors to prevent house-training mishaps. Dogs with mobility impairments may do better with ramps or physical support devices. You may have to assist dogs in areas they previously navigated solo. To help prevent sleep cycle disruptions and maintain appropriate light cycles, give access to natural light and encourage outdoor activities during the day. Also, reduce your dog's exposure to artificial light at night. Play sessions and activity prior to sleep may also enhance your dog's ability to sleep and wake in a natural rhythm as well as reinforce your bond and improve happiness (Landsberg et al. 2013).

As CDS progresses, caregivers should monitor their dog to assure they are maintaining a reasonable quality of life. Recently a scale has been developed to assess the severity and progression of CDS. It is called the CADES canine dementia scale and it measures changes related to behavior of the aging canine.[3] The CADES scale is useful in tracking the progression of symptoms of CDS with mild and moderate cognitive impairment. In a study using the CADES scale to map the development and outcome of CDS, Madari and colleagues (2015) found that social interactions and the sleep–wake cycles were the most impaired functions in elderly dogs. Dogs with CDS showed less frequent or intense interactions with family members and some dogs failed to recognize their caregivers. In the latest stages of CDS dogs showed loss of spatial orientation, as well as a loss of appropriate house-training behaviors. The CADES scale may be useful to caregivers to track the progression of CDS symptoms and can be used for the long-term assessment of the progression of CDS and help determine the best treatment.

Euthanasia decisions and options

Deciding to euthanize your dog is one of the most agonizing tasks you may face as part of your caregiving role. Dogs are an intrinsic part of our everyday lives. They accompany us through our

---

3 The CADES scale describes three stages of cognitive decline in dogs – mild, moderate, and severe cognitive impairment and describes the extent of impairment at each stage (Madari et al. 2015).

daily tasks, sometimes joining us at work, in exercise, and even in social outings. They are so ingrained in our lives that it is hard to imagine going about our routines without them. For me, this was one of the greatest hurdles. Mojo became a Canine Good Citizen[4] and helped me in my work with clients. She joined me on hikes and on runs and kept me company in my office. I had to consider my own desires and needs in coping with no longer having her with me, as well as her need for relief. The human–animal bond is a dynamic relationship, which influences the health and well-being of both parties, and as with all aspects of caregiving we must consider both facets of the relationship. You need to be as prepared as possible for how you will feel about your decision. As a caregiver, you may begin to regret your decision. This regret may lead to feelings of guilt and anxiety if you do not honor your feelings. That is why it is essential to explore your beliefs about euthanasia and come to terms with your role as a caregiver, before your dog's last days arrive. It is also necessary to determine your priorities. Even though there may be options to extend your dog's life, it is not always feasible economically or logistically.

> Ask yourself questions such as:
> Why do I think it is time?
> What are my beliefs about euthanasia?
> What are my fears about the process?
> Whose interests am I taking into account?
> Am I making the decision that is best for me or for my dog?

Consider what is the most important aspect of your pet's end-of-life treatment: Do you want to extend your pet's life at all costs? Is there a point at which you will determine that treatment costs will exceed the benefit your dog will receive from treatment? Do you want to avoid a prolonged deterioration? Often, well-loved pets are humanely allowed to pass in order to alleviate suffering, but it is an individual choice and understanding your priorities will help you make the best decision for you and your dog. The guide "How Do I Know it is Time" from The Ohio State University Veterinary Teaching Hospital helps caregivers explore what is most important in considering a dog's end-of-life treatment and it asks caregivers to reflect on the following:

- Would I consider euthanasia if the following were true about my pet:
    - Feeling pain?
    - Can no longer urinate and/or defecate?
    - Starts to experience seizures?
    - Has become uncontrollably violent or is unsafe to others?
    - Has stopped eating?
    - Is no longer acting normally?
    - Has a condition that will only worsen with time?
    - Financial limitations prohibit treatment?
    - Palliative (hospice) care has been exhausted or is not an option?
    - The veterinary team recommends euthanasia?

Use these questions and the QoL assessments to help guide your decision and always discuss your concerns and pose questions about end-of-life decisions to your veterinarian.

---

4 American Kennel Club CGC Title, http://www.akc.org/dog-owners/training/canine-good-citizen/about/

Veterinarians can be essential to your understanding and coping with the passing of your beloved dog as well as a source of information about the care of pet remains. They can also be a source of support. In a study of caregivers facing the decision to euthanize a pet, 81 percent of clients reported that their veterinarian was a primary source for information about euthanasia and caring for remains. Thirty-three percent of caregivers felt their veterinarian was the contact person to talk to about pet loss (Tzivian et al. 2015). In many cases, euthanasia can be done at your home, if you wish (Box 10.7 lists a few online resources for more information on in-home euthanasia). In addition, Boxes 10.8, 10.9 and 10.10 list other resources for support and grief counseling. Euthanasia and pet loss are very stressful. Studies show that stress levels are higher in bereaved owners, but social support from others including a compassionate veterinarian improved caregivers' reported quality of life (Tzivian et al. 2015). Use the resources provided here as well as the guidance of your support system, including your veterinarian, to bolster you. Remember that the timing may never feel exactly right, but that you can only do what you think is best for you and for your dog. You know your dog better than anyone, and are the right person to make this decision along with advice from your veterinarian, as hard as it may be.

> **In his article "How to Say Goodbye," Andy Roarke, DVM, MS, gently advises:**
>
> *When our pets are suffering, they don't reflect on all the great days they have had before, or ponder what the future will bring. All they know is how they feel today. By considering this perspective, we can see the world more clearly through their eyes. And their eyes are what matter.*
>
> ("How to Say Goodbye" April 16, 2013) http://www.vetstreet.com/our-pet-experts/how-to-say-goodbye?page=2)

Euthanasia for behavior issues

Unfortunately, despite all of the advances in dog training, behavior modification, and psychopharmacological treatment, some dogs are beyond our help. If your dog is exhibiting behavior issues such as aggression and anxiety, seek the help of a veterinary professional to rule out medical causes or contributors to the behavior, and then seek a behavior professional to evaluate your dog and provide treatment with behavior modification (see Chapter 7). When all other treatment options fail, sometimes caregivers are faced with the most difficult decision of all; euthanasia. One of the most painful aspects of behavior consultation is supporting caretakers of truly dangerous or severely emotionally compromised dogs. I have counseled a number of utterly heartbroken caretakers through the process of determining whether or not euthanasia is necessary for the extremely behaviorally compromised pet. Choosing to euthanize your dog is an onerous process. Caretakers experience guilt, self-blame and fear of being criticized or ostracized by their community. No person who loves dogs enough to bring them into their home, and invests the time and resources into medical and behavioral treatment desires euthanasia as the final outcome. Caring for an aggressive dog or a dog with extreme unrelenting anxiety is an arduous, time consuming, financially burdensome, legally compromising, and emotionally draining task.

In these cases, it is sometimes impossible to ensure the caretaker's safety or the dog's safety, let alone well-being. Dogs that are aggressive must be carefully managed. A dog that is biting the caretaker or members of the household must be securely contained, often muzzled, and lives an extremely limited existence. A dog that suffers continually from truly abnormal levels of fear and anxiety must be tightly managed and medicated, to avoid escape or self-injury. We should ask, "What is the quality of this kind of life?" Some feel that euthanasia of an otherwise "healthy" dog with behavior issues is never justified. I would argue that seeing a dog suffer with the distress of their mental disease and the strict management that must be implemented for safety would influence even the most adamant opponent to feel that euthanasia might be the kindest decision for these troubled beings. Although these dogs may be physically healthy, they may not be mentally or behaviorally healthy.

Rehoming a pet with severe behavior issues is almost never a good solution, because a dog with behavior issues is experiencing considerable stress, anxiety, and reduced quality of life. Moving him to a new environment will only cause more distress. Often dogs that have aggression to children are believed to be fine in a home without children, however, this is a misconception. Even though the dog's current target is a child, given the appropriate stressors the dog may show aggression to other targets. Owning a dog with severe, intractable behavior issues is a burden both emotionally and financially, and it is not appropriate to pass this burden along to an unsuspecting new caregiver. If you are feel that you would like to rehome an aggressive or anxious dog, you must be very honest about your dog's behavior and willing to take responsibility.

There are a few sanctuaries that will allow dogs with behavior issues to live out their remaining days. However, this kind of transition and life may be too difficult for some dogs to ever reach a point of proper welfare or well-being. In some sanctuaries, dogs must be kept in individual enclosures without access to other dogs or people because of the potential for aggression. If we consider the Five Freedoms, it is possible to see that the dogs' basic needs for quality of life may not be met in this environment. This position may be somewhat controversial but it is my goal to always defer to the welfare of the dog, and a dog that cannot live out its Five Freedoms is not meeting the criteria for well-being. A dog that lives in a situation in which he feels he must fight or aggress at all times is a very anxiety-ridden dog, and in my view, not a happy dog. Part of our responsibility as caregivers is assuring that those dogs are freed from that stress. In these cases, euthanasia is not a cruel response to an unwanted behavior, but an act of compassion.

Always consider the follow before deciding to euthanize your dog for behavior issues[5]

1. *Rule out medical and physical causes for behavior.* Pain, injury, or illness can result in aggression, especially when there is a sudden onset of symptoms, and in older animals. Seek support from your veterinarian to determine if there could be a medical cause for your animal's behavioral changes.

---

[5] Adapted from: "Behavioral Euthanasia," a handout from The Ohio State University Veterinary Medical Center. Kirby-Madden, T., Shreyer, T., Nielsen, J., & Herron, M. (2014).

2. *Seek professional help.* An academically trained behavior specialist (see Chapter 7 for resources) should be consulted for serious aggression and/or anxiety problems. This may include a veterinary behaviorist (www.dacvb.org) or an applied animal behaviorist (www.certifiedanimalbehaviorist.com). Medications may be helpful for your pet's aggression or anxiety problems. Keep in mind these drugs often take six weeks or more to produce an effect, so make sure to give these treatments enough time before making a decision on further actions.
3. *For dogs that are not a safety risk, consider finding a new home more suited to their needs..* Some problem behaviors may be successfully managed in another setting. For example, a dog with "resource guarding" issues (such as growling over food) may be more safely managed in a home without young children or with a person experienced in dealing with this issue. Dogs with anxiety and fear of children may do better in a quieter home without children. If you decide to rehome your animal you *must* provide your pet's full behavior history to the new owners.

    There is a significant ethical and legal responsibility to ensure your pet does not hurt any people or animals. A safe home would include new owners who are willing to avoid physical and verbal punishment as behavior management, and able to seek help from an academically-trained and experienced behaviorist to better cope with the inappropriate behavior. In general, these homes can be very difficult to find. Animals with behavior problems are often at great risk for abuse and neglect in new environments because they may not have a significant human–animal bond (close relationship with the owner), which is necessary to protect them.
4. *Rehoming is not an option for dogs that bite.* Some animals may not be safe in any environment. For example, a large-breed dog that is aggressive towards strangers will remain a safety concern no matter who adopts him. Most shelters will not adopt out animals with aggression or separation anxiety, and yet you maintain a legal and ethical responsibility to disclose this information.
5. *Environmental factors.* Your living situation may be unchangeable, and most serious problem behaviors require a good deal of environmental modification (avoiding triggers to aggression). Young children and elderly relatives may be more at-risk for bites, and recommended behavioral modification might be extraordinarily difficult or impossible to implement given your living situation. Liability is a concern, as well as the safety of your family and other pets. For example, a dog with aggression towards children cannot be reasonably accommodated in a home with children. Every family has unique circumstances, and you may find that you are just not able to provide the environmental changes that are needed to safely keep your pet.
6. *Suffering.* Animals with behavioral problems have underlying fear, anxiety, and distress, and the owners of these pets often report sharing these feelings as a result. Mental suffering may not be as visible as physical pain, but detracts from your pet's quality of life. Ask yourself: Is my pet having more bad days than good? Can he still enjoy his favorite activities? Is he able to spend time with his people, or does he need to be isolated for safety? Additionally, these pets can create significant stress and anxiety for their human family members. Many people find it helpful to create a quality-of-life log for both themselves and their pet to demonstrate what emotional toll the problem(s) may be causing.

What to expect during the process of euthanasia[6]

Making the decision to say goodbye to a beloved pet is stressful. Having the information about what happens during the visit to the veterinarian can relieve some of the anxiety surrounding the process.

Your veterinarian will generally explain the procedure to you before he or she begins. Do not hesitate to ask your veterinarian for further explanation or clarification if needed. Your veterinarian may ask if you wish to spend a few moments alone with your dog prior to the procedure; if this is not mentioned, you can request it.

Small to medium-size pets are usually placed on a table for the procedure, but larger dogs may be more easily handled on the floor. Regardless of the location, make sure that your dog has a comfortable blanket or bed to lie on.

In most cases, a trained veterinary technician will hold your pet for the procedure. The veterinary technician has the skill needed to properly hold your pet so that the process goes quickly and smoothly. If you plan to be present during the entire procedure, it is important that you allow enough space for the veterinarian and technician to work. Your veterinarian will probably show you where to stand so that your dog can see you and hear your voice.

Your veterinarian will give your pet an overdose of an anesthetic drug called sodium pentobarbital, which quickly causes unconsciousness and then gently stops the heartbeat. Your veterinarian will draw the correct dose of the drug into a syringe and then inject it into a vein.

In dogs, the front leg is most commonly used. The injection itself is not painful to your dog.

Often, veterinarians will place an intravenous (IV) catheter in the pet's vein before giving the injection. The catheter will reduce the risk that the vein will rupture as the drug is injected. If the vein ruptures, then some of the drug may leak out into the leg, and it will not work as quickly.

Your veterinarian may give your pet an injection of anesthetic or sedative before the injection of sodium pentobarbitol. This is most often done in pets that are not likely to hold still for the IV injection. An anesthetic or sedative injection is usually given in the rear leg muscle and will take effect in about five to 10 minutes. Your pet will become very drowsy or unconscious, allowing the veterinarian to more easily perform the IV injection.

Once the IV injection of sodium pentobarbitol is given, your pet will become completely unconscious within a few seconds, and death will occur within a few minutes or less.

Your veterinarian will use a stethoscope to confirm that your pet's heart has stopped.

Your pet may experience some muscle twitching and intermittent breathing for several minutes after death has occurred. Your pet may also release his bladder or bowels. These events are normal and should not be cause for alarm.

After your veterinarian has confirmed that your pet has passed, he or she will usually ask if you would like to have a few final minutes alone with your pet.

---

6 Adapted from the American Humane Association "What to Expect" http://www.americanhumane.org/animals/adoption-pet-care/caring-for-your-pet/euthanasia-decision.html

Burial and cremation options

When your dog leaves this world, it does not mean that he leaves your heart. Rituals are a way to honor and remember our beloved dogs. Your veterinarian can offer you a variety of options for your pet's final resting place (see online resources, Box 10.5). There are a number of pet cemeteries worldwide that can offer you a place to lay your dog to rest.

Cremation is the most popular choice, and you can choose whether or not you would like to have your pet's ashes returned to you. Most cremation services offer a choice of urns and personalized memorials (Box 10.6). Memorials and remembrances may be an important part of your planning for your loss.

Home burial is another option. You may want to bury your pet in your own yard, but before doing so, be sure to check your local ordinances for any restrictions.

Grief and coping with the loss of your dog

First of all, if you have recently had to cope with the loss of a much loved dog, I'm sorry. Many of us have experienced that pain and sorrow of losing a close companion. Experiencing

---

Box 10.5. Pet burial and crematory services

Pet cemetery directory
   http://www.everlifememorials.com/v/pet-loss/pet-cemetery-directory.htm

International Association of Pet Cemeteries
   https://www.iaopc.com/pet-owners/member-directory

Association for Pet Loss and Bereavement-Pet Cemeteries and Crematoria
   http://aplb.org/services/aftercare.php

Association of Private Pet Cemeteries and Crematoria
   http://www.appcc.org.uk

Box 10.6. Pet urns and memorials

http://www.foreverpets.com

http://www.mainelyurns.com/pet-urns.html

http://www.perfectmemorials.com/pet-urns/

http://www.memorialgallerypets.com/dog-urns.aspx

http://rainbowbridgeurns.com/peturns.aspx

http://www.dfordog.co.uk/pet-urns-ashes-keepsakes.html

http://www.legendurn.co.uk/pet-cremation-funeral-urns.html

> ## Box 10.7. In-home euthanasia resources
>
> In-home pet euthanasia directory
> http://www.inhomepeteuthanasia.com
>
> In-home pet euthanasia directory
> http://www.lapoflove.com/Services/In-Home-Euthanasia
>
> A vet's view of in-home euthanasia
> http://www.vetstreet.com/our-pet-experts/a-vets-view-of-home-euthanasia-for-pets
>
> International Association of Animal Hospice and Palliative Care
> https://www.iaahpc.org/resources-and-support/find-help-now.html

grief and mourning the loss of your dog is part of a natural process, but it is a taxing one. Grief is a constellation of thoughts in response to the death and loss of your dog. Mourning is the outward expression of that grief. You may feel grief when your dog is first diagnosed or when their health begins to decline. In a healthy dog with behavior issues, it may begin when you realize that you cannot manage your dog in a safe, humane, and responsible manner. You may feel this loss very deeply, and people that have not experienced such a close bond with a beloved dog may not understand your mourning (and they may even imply you are overreacting). As noted in *When Your Pet Dies: A guide to mourning, remembering, and healing* (Wolfelt 2004), your feelings are what they are and you are having them for a reason. Mourning (expression of grief) is a necessary process for most people to move forward.

Research suggests that people cope best with the decision to euthanize their dog if the following have occurred (Kay et al 1988):

1. Discussion of the dog's condition, prognosis, expected costs of care, and expected cause of death (if not euthanasia).
2. Understanding of the method, reason for, and cause of death.
3. Expected behaviors and feelings of the dog during death.
4. Who makes the decision and if they have support (individual, couple, or family support).
5. Time to prepare for the death of the dog.
6. Understanding of the procedure and assurance that the death will be peaceful.
7. Time with the dog either before or during death.
8. Reassurance that it is OK to express grief, cry, and regain composure before having to leave the dog.
9. Having a realistic understanding of the life span and life cycle of the dog.
10. Freedom of guilt for acts of omission or commission.
11. Willingness to participate in support groups or seek the support of friends, family, and others that allow for bereavement activities.
12. Having knowledge that the veterinarian and the behavior professional have done what is possible for that condition, within the ability to bear the costs and for the quality of life for your dog.

> **Box 10.8. Recommended reading for more information on coping with the loss of your dog**
>
> *The Loss of a Pet: A guide to coping with the grieving process when a pet dies*, by Wallace Sife, Howell Book House, 2014
>
> *The Grief Recovery Handbook for Pet Loss*, by Russell Friedman, Cole James, and John W. James, Taylor Trade Publishing, 2014
>
> *Grieving the Death of a Pet*, by Dr. Betty Carmack, R.N., Ed.D., Augsburg Publishers, 2003
>
> *Pet Loss: A thoughtful guide for adults and children*, by Herbert A. Nieburg and Arlene Fischer, Harper & Row, 1982
>
> *Coping With Sorrow On The Loss of Your Pet*, by Moira Anderson, M.Ed, Alpine Blue Ribbon Books, 1996
>
> *The Grief Recovery Handbook for Pet Loss*, by Russell Friedman, Cole James, and John W. James, Taylor Trade Publishing, 2014

13. Having purposeful remembrance and recognition by others including your veterinarian, behavior professional, and others about your deceased dog.
14. Having knowledge that there is counseling for the caregiver and assistance coping with grief and moving on after the loss of your canine companion.

Guilt and regret and self-blame are common emotions when dealing with pet loss.

These feelings are natural and generally represent the need to express your grief. Find understanding supportive people that are willing to listen. Sadness, loneliness, and depression are natural reactions to the loss of your companion. Allowing yourself to feel sorrow is a part of the grieving and healing process. Allow yourself to experience these emotions, but be aware of the difference between sadness and depression. While sadness is normal and healthy, clinical depression is an illness that needs to be addressed. If you feel like your grief is turning into untreated clinical depression,[7] please seek the advice of your health or mental health care provider. There are a number of pet loss support groups as well as pet loss hotlines to assist you in moving through your grief and mourning process (Box 10.10).

Helping children and adolescents to cope with the loss

Dogs are a part of our family lives and it is a sad reality that children within these families will have to cope with the loss of a companion dog. You may find yourself at a loss for words, not knowing how to share information about the death or how to best help children and adolescents cope with their grief.

---

[7] There are a number of resources for recognizing depression, you can find one example at http://www.webmd.com/depression/symptoms-depressed-anxiety-12/

There are some important things to considering when your family is faced with this challenge.

1. Your child may have very deep emotions about your dogs. They may even seem to value the life of the dog over even extended relatives. It may be difficult to hear comments to this effect, or that they appear more upset over the loss of their dog than they appeared to be when they suffered the loss of a relative. It is important that you remember that this emotional intensity is normal.
2. Do not minimize the emotions of your child. Some people may comment, "It was only a dog." Understand, to your child (as is true for many adults), the dog may have been a best friend.
3. Help your child through their grief; do not ignore it. There are resources that can help children and teens cope with the loss (Boxes 10.9 and 10.10). Watch for signs that the child is not recovering from their grief and if necessary, get them the help that they need from a health care provider.

Telling your child about the loss of your pet is an emotionally wrought topic. Choose your words thoughtfully and wisely. Children are very literal and may find it scary if you use words like the dog was "put to sleep" or "he was sick so we had to put him to sleep." Children may imagine that they might not awaken when they fall asleep, and could develop sleep issues. They might imagine that such a fate might befall them, if they too were to become ill. Tell your child in words that they can understand, such as this:

> Fritz was old and ill. Even though we loved him very much, he was not able to get better. His body was too sick. So we took him to the doctor and asked the doctor to give Fritz some medicine that would help him to be more comfortable while his body was failing. Mom stayed with Fritz while Fritz died, and now Fritz is running around in heaven (or use reassurances appropriate to your belief system) doing all of the things he did when he wasn't old and sick.
> (Pelar 2013)

Tell your child the truth in a way that is appropriate to his age. If you mislead your children, the truth usually comes to light and leaves the child feeling a sense of mistrust and betrayal. Avoid telling children that the dog ran away or went to live on a farm. This may leave children wondering why you have not made an effort to find him, or that you sent him away.

Children may experience a variety of emotions in reaction to the loss of a dog. They may experience sadness, loneliness, guilt, frustration, and even anger. Help children understand that it's natural to feel all of those emotions, and reassure them that they can talk to you when they are ready. Children may or may not want to share their feelings at first, or even hear about your grief. It is OK to show that you are sad and to share your stories and memories, but be considerate and respectful of their reactions. Remember too that you are a role model. You might not feel like it at the time, but you are teaching your child about dealing with loss through your own behavior. Be mindful of what you are experiencing and how that may translate to your child. Tell your child that their pain is valid, but that it will fade over time; there will likely be happy days and sad days in the weeks or months that follow. Grief is not always a linear process; it may take on many forms, and your child may appear fine one day and then flooded with emotion the

> **Box 10.9. Recommended books for children**
>
> *Helping Children to Cope with Separation & Loss*, by Claudia Jewitt, Harvard Common, 1992
>
> *Snowflake in My Hand*, by Samantha Mooney, Delacorte, 1983
>
> *The Tenth Good Thing About Barney*, by Judith Viorst, Atheneum, 1975
>
> *When A Pet Dies*, by Fred Rogers, G.P. Putnam's Sons, 1988

next. Help them to see that there is hope for happier times and encourage them to focus on how much joy their companion brought to the family. Explain to children that dogs have shorter lives than ours and despite the pain that losing them brings, the happy lives we give to dogs and the happiness they give us in return make up for it, many times over.

It can help kids to find special ways to remember a beloved dog. You might have a ceremony to bury him or his ashes, or just share memories of fun times you had together. Write a remembrance together or offer thoughts on what the pet meant to each family member. Share stories of your pet's funny moments or escapades. You could do a project, too, like making a scrapbook. Perhaps most importantly, talk about your pet, often and with love. When the time is right, you might consider adopting a new dog — not as a replacement, but as a way to welcome another animal friend into your family.

Healing and moving on after the loss

According to Alan D. Wolfelt in his guide, *When Your Pet Dies*, when a beloved dog dies, caretakers have six needs to reconcile grief and move toward healing.

1. *Acknowledge the reality of the death*: Whether the loss is sudden or anticipated, the full reaction to the loss may take weeks or months. Going through daily routines like coming home without the standard greeting at the door or settling down to read without your companion next to you is a gradual adjustment. Getting used to the idea that your dog is gone will take time.
2. *Move toward the pain of the loss*. Experience and explore your thoughts and feelings about the death of your beloved dog. You may find that you busy yourself with distractions or avoid activities or places that remind you of your dog. While you should not try to confront all painful feelings at once, do allow yourself to be sad and work through those feelings.
3. *Continue to honor your dog through memories*. As you begin to recover from the loss, embrace your memories (both happy and sad). Remembering the past makes hoping for the future possible.
4. *Adjust your self-identity*: Part of your self-identity comes from being a dog caretaker. Others see you that way too. Perhaps you are the person with the big goofy lab or the greeter at

The Rainbow Bridge

Just this side of heaven is a place called Rainbow Bridge. When an animal dies that has been especially close to someone here, that pet goes to Rainbow Bridge. There are meadows and hills for all of our special friends so they can run and play together. There is plenty of food, water and sunshine, and our friends are warm and comfortable.

All the animals that had been ill and old are restored to health and vigor. Those who were hurt or maimed are made whole and strong again, just as we remember them in our dreams of days and times gone by. The animals are happy and content, except for one small thing; they each miss someone very special to them, who had to be left behind.

They all run and play together, but the day comes when one suddenly stops and looks into the distance. His bright eyes are intent. His eager body quivers. Suddenly he begins to run from the group, flying over the green grass, his legs carrying him faster and faster. You have been spotted, and when you and your special friend finally meet, you cling together in joyous reunion, never to be parted again. The happy kisses rain upon your face; your hands again caress the beloved head, and you look once more into the trusting eyes of your pet, so long gone from your life but never absent from your heart.

Then you cross Rainbow Bridge together …

Paul C. Dahm. 1998. *The Rainbow Bridge*. Turning Side Press.

the dog park. Now that your dog has died, this part of your identity has changed. You need to recognize that your role has changed and adjusting to this change may be difficult but is also a part of the healing process.

5. *Search for meaning*: When a beloved dog dies we may question the meaning and purpose they play in our lives. Why do we have them when their lives are so short? Will I ever be able to love another dog like I loved this one? Coming to terms with the answers to these questions at your own pace will allow you to determine what is important to you. Asking yourself to think through these questions is an important exercise in mourning and healing, even if there are no concrete answers.
6. *Continue to receive support from others*: As you learn to cope with the death of your pet, you may need help less intensely and support less often, but you will always need friends and family to listen and provide encouragement. Talking to others who have experienced similar loss may help. Seek pet loss support groups (Box 10.10), pet loss help lines or other pet caretakers to help guide you through your mourning. Although things do get better with time, mourning is an active process. Work through your emotions and take care of yourself so that you can move on, not forgetting but by honoring your dog with a happy life filled with memories of your friend.

Box 10.10. Pet loss support on-line resources and pet loss hotlines

http//www.pet-loss.net (The Pet Loss Support Page)
http://www.aplb.org (The Association for Pet Loss and Bereavement)
http://www.petloss.com (Pet Loss Grief Support)
http://www.rainbowbridge.com (The Rainbow Bridge, A Pet Loss Community)

Washington State University College of Veterinary Medicine
Pet Loss Hotline
1-(866) 266-8635 or (509) 335-5704
http://www.vetmed.wsu.edu/outreach/pet-loss-hotline

Cornell University College of Veterinary Medicine
Pet Loss Support Hotline
607-253-3932
6:00pm–9:00pm Eastern Time
Every Monday, Wednesday, Thursday, and Saturday

Tufts University Pet Loss Support Hotline
508-839-7966
http://vet.tufts.edu/petloss/

University of Illinois
Toll-free: 877-394-2273 (CARE)
www.cvm.uiuc.edu/CARE/

Michigan State University
517-432-2696
cvm.msu.edu/alumni-friends/information-for-animalowners/pet-loss-support

In the UK
Blue Cross for Pets
088 096 6606
https://www.bluecross.org.uk/pet-bereavement-support

Animal Samaritans Pet Bereavement Service
020 8303 1859
www.animalsamaritans.org.uk

EASE Pet Loss Support Services
www.ease-animals.org.uk offer a wide range of support resources, covering different aspects of grief in pet bereavement; free to download or view or listen to online (face-to-face support is available by arrangement in East Devon only).

Pet Bereavement Support Service
0300 777 1897
www.bluecross.org.uk

How do I know if it is time to get a new dog?

I get asked often when is the best time to add a new dog after the loss of a dog. The decision varies for each individual but should be carefully considered. It is important that you do not rush into a decision on a new companion, simply to comfort you in your loss. Explore your motivations; you may have an impulse to adopt or purchase a new dog simply because you have not fully adjusted to the loss of your departed pet. In this scenario, many people end up disappointed that their new dog is not the same as the one they have lost. To best discover if the time is right and to find the next special dog, consider the following:

1. Do not try to replace your old dog. It is OK to look for the qualities of your previous dog that you admired in your new dog but understand that your new dog has its own personality traits and behaviors. Understand that your relationship with your new dog will take time to develop.
2. Understand that a new dog might take more time and effort than your previous dog and determine if you have the time to devote to your new companion.
3. Make sure that if there are multiple people in the household that all opinions are considered. Everyone processes the loss of a pet differently and although you might feel ready, the addition of a new dog may feel like a betrayal or cause other hurt feelings in others. Make sure everyone is in agreement to bring home a new dog.
4. Consider volunteering at a shelter or fostering a dog from a rescue as an entry back into dog caregiving. You may find that you are not ready for the amount of work required to care for a dog or that you still have not reconciled with your loss. You may also surprise yourself by bonding with a special dog in need and provide a loving home to a beloved shelter pet.
5. Reeducate yourself! Perhaps it has been a while since you raised a new puppy or started training with a new dog. There are a number of advances in dog training and enrichment (see Chapter 5 for training tips and Chapter 6 for enrichment ideas). The best way to ensure success with your new dog is to be prepared both emotionally and educationally. Someday soon you will be ready to have another happy dog!

References and additional resources

Adams, C.L., Bonnett, B.N., & Meek, A.H. (1999). Owner response to companion animal death: Development of a theory and practical implications. *The Canadian Veterinary Journal* 40: 33–39.

Belshaw, Z., Asher, L., Harvey, N.D., & Dean, R.S. (2015). Quality of life assessment in domestic dogs: An evidence-based rapid review. *Veterinary Journal* 206: 203–212.

Brown, D.C., Boston, R.C., Coyne, J.C., & Farrar, J.T. (2007). Development and psychometric testing of an instrument designed to measure chronic pain in dogs with osteoarthritis. *American Journal of Veterinary Research* 68: 631–637.

Brown, D.C., Boston, R.C., Coyne, J.C., Farrar, J.T. (2008). Ability of the Canine Brief Pain Inventory to detect response to treatment in dogs with osteoarthritis. *Journal of the American Veterinary Medical Association* 233: 1278–1283.

Brown, D.C., Boston, R., Coyne, J.C., & Farrar, J.T. (2009). A novel approach to the use of animals in studies of pain: Validation of the Canine Brief Pain Inventory in canine bone cancer. *Pain Medicine* 10: 133–142.

Dahm, P.C. (1997). The Rainbow Bridge. Oceanside, OR: Running Tide Press.

Dodd, C.E., Zicker, S.C., Jewell, D.E., et al. (2003). Can a fortified food affect behavioral manifestations of age-related cognitive decline in dogs? *Veterinary Medicine* 98: 396–408.

Fernandez-Mehler, P., Gloor, P., Sager, E., et al. (2013). Veterinarians' role for pet owners facing pet loss. *Veterinary Record* 172: 555.

Freeman, L.M., Rush, J.E., Farabaugh, A.E., & Must, A. (2005). Development and evaluation of a questionnaire for assessing health-related quality of life in dogs with cardiac disease. *Journal of the American Veterinary Medical Association* 226: 1864–1868.

Kay, W., Cohen, S., Fudin, C., Kutcher, A., Nieburg, H., Grey, R., & Osman, M. (Eds.). (1988). *Euthanasia of the Companion Animal: The impact on pet owners, veterinarians, and society*. Philadelphia, PA: The Charles Press. 267 p.

Kirby-Madden, T., Shreyer, T., Nielsen, J., & Herron, M. (2014). Euthanasia for behavior issues: A complicated and difficult decision. https://vet.osu.edu/vmc/sites/default/files/import/files/documents/pdf/vmc/Behavioral%20Euthanasia%20fact%20sheet.pdf

Landsberg, G.M., Nichol, J., & Araujo, J.A. (2012). Cognitive dysfunction syndrome: a disease of canine and feline brain aging. *Veterinary Clinics of North America. Small Animal Practice* 42: 749–768.

Landsberg, G.M., Hunthausen, W., & Ackerman L. (2013). The effects of aging on behavior in senior pets. In *Handbook of Behavior Problems of the Dog and Cat*. 3rd edition. Philadelphia: WB Saunders. pp 211–235.

Lynch, S.J., Savary-Bataille, K., Leeuw, B., & Argyle, D.J. (2011). Development of a questionnaire assessing health related quality of life in dogs and cats with cancer. *Veterinary & Comparative Oncology* 9: 172–182.

Madari, A., Farbakova, J., Katina, S., et al. (2015). Assessment of severity and progression of canine cognitive dysfunction syndrome using the Canine Dementia scale (CADES). *Applied Animal Behavior Science* 171: 138–145.

Nielson, J. (2013). How Do I Know it is Time? The Ohio State University Veterinary Medical Center. https://vet.osu.edu/vmc/sites/default/files/import/assets/pdf/hospital/companionAnimals/HonoringtheBond/HowDoIKnowWhen.pdf

Osella, M.C., Re, G., Odore, R., et al. (2007). Canine cognitive dysfunction syndrome: Prevalence, clinical signs and treatment with a neuroprotective nutraceutical. *Applied Animal Behaviour Science* 105: 297–310.

Pelar, C. (2013). *Living with Kids and Dogs Without Losing Your Mind: A parent's guide to controlling the chaos*. Woodbridge, VA: Dream Dog Productions. 184 p.

Reid, J., Wiseman-Orr, M.L., Scott, E.M., & Nolan, A.M. (2013). Development, validation and reliability of a web-based questionnaire to measure health-related quality of life in dogs. *Journal of Small Animal Practice* 54: 227–233.

Tzivian, L., Friger, M., & Kushnir, T. (2015). Associations between stress and quality of life: Differences between owners keeping a living dog or losing a dog by euthanasia. *PLoS ONE* 10.1371

Wolfelt, A.D. (2004). When Your Pet Dies: A guide to mourning, remembering, and healing. Fort Collins, CO: Center for Loss of Life Transition.

Web resources

Association of Pet Loss and Bereavement, http://www.aplb.org/resources/quality-of-life_scale.php – Pet quality of life rating scale, based on an original concept from Villalobos, A.E. (2004) Quality of life scale helps make final call, *Veterinary Practice News*, 09/2004, and included in Dr. Villalobos' 2006 book, *Canine and Feline Geriatric Oncology: Honoring the human–animal bond*, Blackwell Publishing.

Annotated Bibliography: a brief list of some of our favorite dog resources

Photo: iStock

ACH INDIVIDUAL CHAPTER OF THIS BOOK CONTAINS A MORE EXTENSIVE LIST OF REFerences for the content included in the book, for those interested in reading the original research and/or learning more about a given aspect of dog science. What follows is a short list of some of our favorite, more general resources for learning about dogs, sorted by subject and chapter. But, this isn't an exhaustive list! There are many more great resources out there than we could fit into our "short list." Friends, students, clients, and colleagues often ask us for our recommendations for further reading about dogs, and so these are the books that we tend to recommend first. Enjoy!

Chapter 1: The human–animal bond, and evolution of the domestic dog

Coppinger, R. and Coppinger, L. (2016) What Is a Dog? Chicago: The University of Chicago Press.
The Coppingers' book is a fascinating take on the evolution of the domestic dog, written more from a perspective like that of a wildlife biologist than many of the dog books on the market today.

Donaldson, J. (1996) The Culture Clash: A revolutionary new way of understanding the relationship between humans and domestic dogs. Berkeley, CA: James & Kenneth Publishers.
Donaldson is a very experienced trainer and writer who has worked extensively with the world famous behaviorist Dr. Ian Dunbar, and her insight into the minds of dogs shines through in this book, which really was (at the time of publication) a bit of a revolution in how we view our dogs. This is another book which I love because of its focus on love and respect for dogs *as dogs* – to truly provide them with the best quality of life, we need to understand and respect their canine nature; our failure to do this can have disastrous outcomes.

McConnell, P. (2002) The Other End of the Leash: Why we do what we do around dogs. New York: Ballantyne Books.
This book focuses on our relationship with our canine companions, with the goal of helping us to understand why dogs behave in the ways that they do (and, how our human behaviors may influence our dogs). Essential reading for improving your communication with your dog, this book is still one of my absolute favorite books about dogs!

McConnell, P. (2007) For the Love of a Dog: Understanding emotion in you and your best friend. New York, NY: Ballantyne Books.
Dr. McConnell's writing is always enjoyable and informative, and this is a lovely book about emotions in dogs and our relationship with them. This book also includes a very extensive annotative bibliography – worth a look if you have already exhausted all the resources on our list and are hungry for more.

Miklosi, A. (2007) Dog Behaviour, Evolution and Cognition. Oxford, UK: Oxford University Press.
Dr. Miklosi was part of one of the first research programs to focus on the domestic dog, working with Dr. Vilmos Csanyi. Written like a textbook, with ample references to the scientific literature, Dr. Miklosi's book is not an easy read like the McConnell books, but for those truly dedicated to understanding the science of the domestic dog, this is a great reference.

Olmert, M. (2009) Made For Each Other: The biology of the human–animal bond. Cambridge, MA: De Capo Press.
Another interesting read on the origins and nature of the human–animal bond, well supported by references to recent science. Not just limited to dogs, Olmert's book looks at the process and outcomes of domestication in cats, horses, livestock, and other species.

Chapter 2: Quality of life in dogs

Association of Shelter Veterinarians (ASV) (2010) Guidelines for standards of care in animal shelters. Newbery, S. et al., editors. Available online (free) at: http://www.sheltervet.org/
This is a technical manual, not a book per se, but very useful reading for anyone working with dogs housed in a kennels or shelter situation. The guidelines provide clear, concise recommendations on everything from behavioral welfare to cleaning to recommended staff ratios, and more.

Bradshaw, J. (2011) Dog Sense: How the new science of dog behavior can make you a better friend to your pet. New York: Basic Books.
A wonderful book in support of the appreciation of dogs as dogs (rather than as wolves, or small furry humans), Dr. Bradshaw's book contains a wealth of practical information on improving the life of your dog.

Grandin, T. and Johnson, C. (2009) Animals Make Us Human: Creating the best life for animals. New York: Hougton Mifflin Harcourt.
This is a remarkable book, and provides great insight into the emotional needs of animals (not just limited to dogs). Dr. Grandin bases her work on her many years as truly groundbreaking scientist in the field of animal behavior and welfare. A must-read for those interested in the emotional lives of animals.

Chapter 3: Canine cognition

Hare, B. and Woods, V. (2013) The Genius of Dogs: How dogs are smarter than you think. New York, NY: Plume Books. 367 p.
This book offers a great summary of the work done in canine cognition at Duke Canine Cognition Lab and other research facilities. It explores the new discoveries about the dog's unique social cognitive abilities.

Horowitz, A. (2009) Inside of a Dog: What dogs see, smell, and know. New York, NY: Scribner.
A somewhat whimsical account of what dogs know, think, and perceive. This book offers a nice introduction to the science of dog cognition.

Horowitz, A. (ed.) (2014) Domestic Dog Cognition and Behavior: The scientific study of Canis familiaris. Heidelberg: Springer.
A good resource for those looking for an in-depth scientific exploration of the current research in dog cognition. Contributors to this book provide innovative information such as how dogs perceive the world through their senses and how different breeds of dogs differ in their behavior.

Kaminski, J. and Marshall-Pescini, S. (Eds.) (2014) The Social Dog: Behaviour and cognition. San Diego, CA: Academic Press.

The Social Dog: Behaviour and cognition is an excellent review of the current research in dog cognition. It covers a vast amount of information about our canine companions on topics such as the history and evolution of the domestic dog, dog cognition and perception, and the dog–human bond.

Chapter 4: Reading your dog

Aloff, B. (2005) Canine Body Language: A photographic guide. Wenatchee, WA: Dogwise Publishing.
A great resource for photographs that illustrate common dog body language. Written in a conversational style, it is a good introduction to dog communication although some of the interpretations of dog communication are lacking scientific support.

Abrantes, R. (1997) Dog Language: An encyclopedia of canine behaviour. Naperville, IL: Wakan Tanka Publishers.
A classic encyclopedia that contains detailed illustrations of canine facial expressions and ethological descriptions of an aggregate of canine behaviors.

Darwin, C. R. (1872) The Expression of the Emotions in Man and Animals. London: John Murray.
A seminal work that provides the foundations to the understanding of canine communication and emotions. This book sets the groundwork for current research discoveries in canine cognition and offers brilliant illustrations of Darwin's account of emotions in dogs.

Miklósi, Á. (2007) Dog Behaviour, Evolution, and Cognition. Oxford, UK: Oxford University Press.
An evolutionary perspective of dog behavior and cognition. This book provides information about attachment behavior between humans and dogs as well as describing the importance of interspecific communication to the success of dogs in human communities.

Scott, J.P. and Fuller, J.L. (1965) Genetics and the Social Behavior of the Dog. Chicago: University of Chicago Press.
This landmark text examines the genetic components of social behavior in dogs. This work illuminates breed-specific differences and the role genetics plays in behavior.

Serpell, J. (Ed). (1995) The Domestic Dog: Its Evolution, Behavior and Interactions with People. Cambridge: Cambridge University Press.
This is a comprehensive review of a number of topics on dogs' behavior and their relationships with humans. It offers detailed explanations of the evolution of modern domesticated dog breeds, dog behavior, and the underlying genetic complexity.

Chapter 5: Dog training

Burch, M.R. and Bailey, J.S. (1999) How Dogs Learn. Hoboken, NJ: Howell Book House.
Burch and Bailey's book is an introduction to operant conditioning and the science of how dogs (and, many animals) learn, and as such is essential background reading for those serious about training dogs. The book is organized in a very easy-to-read format, with lots of helpful diagrams and practical tips.

Donaldson, J. (2010) Train Your Dog Like a Pro. Hoboken, NJ: Howell Book House.
An excellent, clear, step-by-step approach (complete with included DVD) to training basic life skills to your canine companion. Donaldson has extensive experience training dogs, and training trainers to train dogs, and it shows in this book and DVD.

McConnell, P. and Skidmore, B. (2010) The Puppy Primer (2nd edition). Black Earth, WI: McConnell Publishing, Ltd.
I would recommend anything Dr. McConnell has written, given her experience, depth of knowledge, and obvious love of dogs. This book is no exception. A nice, concise, and easy-to-follow guide to raising a puppy, including training tips and troubleshooting issues.

Miller, P. (2008) The Power of Positive Dog Training (2nd edition). Hoboken, NJ: Howell Book House.
Miller's book is a comprehensive users' guide, with step-by-step instructions, to rewards-based dog training. More essential reading for those serious about training dogs, using the most humane and science-based approaches!

Tillman, P. (2000) Clicking With Your Dog: Step-by-step in pictures. Waltham, MA: Karen Pryor Clickertraining/Sunshine Books.
An excellent, user-friendly introduction to clicker training, including instructions on how to train a wide variety of canine life skills: sit, stay, come when called, crate training, leave it, loose-leash walking … it's in here, and much more.

Yin, S. (2010) How to Behave so Your Dog Behaves. (2nd edition). Neptune City, NJ: T.F.H. Publications, Inc.
A great, fun, introductory book for those new to owning or living with a dog, or for those interested in learning about training their dogs using training methods based on the science of how dogs learn. Also includes an extensive section on addressing common behavior problems in dogs.

Chapter 6: Canine mental health and enrichment

Braitman, L. (2014) Animal Madness: Inside their minds. New York: Simon and Schuster.
Animal Madness is a lovely and fascinating book about having empathy and understanding for the inner emotional lives of our fellow creatures. The book incorporates scientific research on

dogs, primates, elephants, and more, into a remarkable and informative glimpse into the inner workings of non-human animals.

Morell, V. (2013) Animal Wise: how we know animals think and feel. New York: Broadway Books.
Like Braitman's book (above), *Animal Wise* is a fascinating (and moving) look into the minds and emotions of animals, and the researchers who have helped to shed light on these inner worlds. Morell is an engaging writer, and this book is a pleasure to read.

Wells, D.L. (2004) A review of environmental enrichment for kennelled dogs, Canis familiaris. Applied Animal Behavior Science 85: 307–317.
This is a scientific article, not a book, but for those interested in learning more about canine enrichment, Wells provides a helpful overview of types of enrichment used with kenneled dogs (from human contact to toys to music to social time with other dogs), and the relative benefits/drawbacks of each.

Chapter 7: Canine behavior problems and solutions

British Small Animal Veterinary Association (BSAVA) (2012) BSAVA Manual of Canine and Feline Behavioural Medicine (2nd edition). Horwitz, D. and Mills, D. (Eds.) Gloucester, UK: BSAVA.
Written for veterinarians, this is a comprehensive and in-depth guide to understanding and treating behavior problems in dogs and cats. Chapters cover topics ranging from basic requirements for good behavioral health in dogs (with a separate chapter for cats), training protocols, preventing behavior problems, house soiling, aggression, and more. This book is a great resource, with each chapter written by a well-respected expert on that particular topic.

Hart, B., Hart, L., and Bain, M. (2006) Canine and Feline Behavior Therapy (2nd ed.) Oxford, UK: Blackwell Publishing.
This is a textbook, but a readable one, which presents an in-depth overview of companion animal behavior, including extensive references to the scientific literature, as well as practical suggestions for resolving behavioral problems.

Horwitz, D. and Neilson, J. (2007) Blackwell's Five-Minute Veterinary Consult: Canine and feline behavior. Oxford, UK: Blackwell Publishing.
Also written for veterinarians, this is the book I recommended most frequently to my veterinary students as the most important reference book to have on their shelf when dealing with behavior problems in their patients. Without meaning to underestimate the seriousness of this topic, this book is like a "cookbook" for treating behavior problems in dogs and cats, with each chapter addressing a specific behavior problem, contributing and risk factors for developing each problem, useful diagnostic tests to employ, and treatment approaches.

McConnell, P. (2005) The Cautious Canine: How to help dogs conquer their fears (2nd edition). Black Earth, WI: McConnell Publishing Ltd.

A short (only 29 pages), inexpensive, easy-to-read, step-by-step, and highly effective guide to treating fear-based behavioral problems in dogs, using counterconditioning. This is only one of Dr. McConnell's very useful booklets on treating various behavioral problems in dogs; check out her website (www.patriciamcconnell.com) for a full list.

The American College of Veterinary Behaviorists (2014) Decoding your dog. Horwitz, D., Ciribassi, J. and Dale, S. (Eds.) New York: Houghton Mifflin Harcourt.
Written by members of the American College of Veterinary Behavior (veterinarians with special certification in treating behavior problems in domestic animals), this is a very practical, user-friendly book on understanding, treating, and preventing behavior problems in companion dogs. Topics covered include aggression, separation anxiety, noise phobias, and more.

Chapter 8: Canine physical wellness

Case, L.P. (2014) Dog Food Logic. Wenatchee, WA: Dogwise Publishing.
This book contains a scientific review of dog nutrition, feeding recommendations, and exercise. Written in a concise easy-to-understand style, the author informs the reader how to take a science-minded approach to decisions about their dog's nutrition.

Dodman, H. (2009) The Well-Adjusted Dog: Dr. Dodman's 7 steps to lifelong health and happiness for your best friend. New York, NY: Houghton Mifflin Company.
Dodman provides a comprehensive approach to the health, behavior, and environment for the life of your dog. He describes how to set the groundwork through proper care and training to provide a happy life for your dog.

Chapter 9: Veterinary care

Bartges, J. et al. (2012) American Animal Hospital Association (AAHA) Canine Life Stage Guidelines. Available online (free) at www.aaha.org/professional/resources/canine_guidelines_abstract.aspx
Written for veterinarians, this document is helpful in understanding your dog's medical needs from the veterinarian's perspective, and may be helpful in navigating your way through various veterinary visits and care needs.

If you are in the process of trying to locate a good veterinarian for your dog, it is worth referring to one of the relevant web pages on this topic. The American Veterinary Medical Association (AVMA) has a page on how to choose a veterinarian (https://www.avma.org/public/YourVet/Pages/finding-a-veterinarian.aspx), as do the British Veterinary Association (https://www.bva.co.uk/You-and-your-vet/Choosing-a-vet/) and the Humane Society of the United States (http://www.humanesociety.org/animals/resources/tips/choosing_a_veterinarian.html).

We strongly recommend seeking out a veterinary practice that uses low-stress handling techniques for the well-being of their patients; for more on low-stress handling, you can check out Sophia Yin's excellent manual:

Yin, S. (2009) Low Stress Handling, Restraint and Behavior Modification of Dogs & Cats. Davis, CA: Cattledog Publishing.
This book is a technical manual written for the veterinary market, but does include useful introductory chapters on the science behind and importance of low-stress handling techniques. For more general information, see Dr. Yin's website, *https://drsophiayin.com/low-stress-handling/*

Chapter 10: End-of-life care and pet loss support

Friedman, R., James, C. and James, J.W. (2014) The Grief Recovery Handbook for Pet Loss. Lanham, MD: Taylor Trade Publishing. 168 p.
A guide that helps caregivers navigate through the grieving processes. This guide pays special attention to the human–animal bond.

Jewett, C. (1992) Helping Children to Cope with Separation & Loss. Harvard: Harvard Common.
A great book for anyone that wishes to gain an understanding of the emotions of a child experiencing the loss of their pet. This book offers warm advice, specific techniques, and innovative ideas for helping children overcome the sadness, anger, and anxiety they feel during a difficult time.

Pelar, C. (2013) Living With Kids and Dogs Without Losing Your Mind: A Parent's Guide to Controlling the Chaos. Woodbridge, VA: Dream Dog Productions. 184 p.
This is a great resource for anyone living with kids and dogs. This book offers a caring and practical guide for helping children cope with their understanding of death and grieving and to move on after the loss of a pet.

Sife, W. (2014) The Loss of a Pet: A guide to coping with the grieving process when a pet dies. Hoboken, NJ: Howell Book House
Written by a psychotherapist and the founder of the Association for Pet Bereavement, Sife provides a guide to coping with grief from losing a pet that might not be understood or respected. It is a compassionate, practical guide through the grieving process.

Wolfelt, A.D. (2004) When Your Pet Dies: A guide to mourning, remembering, and healing. Fort Collins, CO: Center for Loss of Life Transition.
This books gives dog caregivers a deeper understanding of the special circumstances surrounding the grief from losing a beloved companion animal. Explains how to cope when others do not understand the impact of losing a dog and the grief surrounding that loss.

Index

AAFCO *see* Association of American Feed Control officials
AAHA *see* American Animal Hospital Association
ACVB *see* American College of Veterinary Behavior
Adaptil 122
adaption hypothesis 33
aggression 107–8, 137, 140–3
   avoid physical reprimands (positive punishment) 145
   avoid situations/triggers that provoke 145
   behavioral context 58
   body posture 64
   books, DVDs, websites 142, 148
   castration 143
   dealing with 143–5
   in the ears 57, 61
   in the eyes 57, 58
   fear as component of 155–6
   in the mouth 57, 60
   muzzles 143–4
   no interaction with animal 144–5
   safety plan 143
   seek help sooner rather than later 143
   socialization, importance of, in reducing risk of 141, 149
   in the tail 63
   threat displays 67–8
   treating 145–9
   warnings should be heeded 144
AHVMA *see* American Holistic Veterinary Medical Association
alpha roll 95, 102, 145
American Animal Hospital Association (AAHA) 168, 197
American College of Veterinary Behavior (ACVB) 128
American Holistic Veterinary Medical Association (AHVMA) 219
American Pet Products Association (APPA) 170
American Veterinary Medical Association (AVMA) 2, 212
American Veterinary Society of Animal Behavior (AVSAB) 94, 128
Animal Behavior Society 128
Animals Make Us Human (2010) 19
anthropomorphism 13–14
APDT *see* Association of Pet Dog Trainers
APPA *see* American Pet Products Association
Association of American Feed Control officials (AAFCO) 183, 185
Association of Pet Dog Trainers (APDT) 82, 101
Association of Shelter Veterinarians (ASV) 108–9
American Association of Zoos and Aquarium's Behavior Scientific Advisory Group 109
associative learning 130
   classical conditioning 75–8, 130–1
   operant conditioning 75, 78–80, 131–2
ASV *see* Association of Shelter Veterinarians
AVSAB *see* American Veterinary Society of Animal Behavior
aversives, use of in training 131 (*see also* punishment, positive, risks of)

Backman, Maureen 144
Bailey, Bob 80, 81
barking 16, 22, 65, 67, 73–4, 91, 121, 132, 133, 137, 139, 151, 156, 158, 159, 226
Beaver, Bonnie 215
Becker, Marty 215
behavior 19
   learning new behavior 86–7
behavior adjustment training (BAT) 142
behavioral problems 127–8
   aggression 137, 140–9
   books/online resources 129
   common 128–30, 137
   counterconditioning 130, 134–6, 139, 148, 150–3, 155–6
   expert help contacts 128
   fear or aggression, distinguishing between 155–6
   fears, phobias, anxiety 137, 149–55
   nuisance or normal behaviors 137–40
   physical rather than behavioral 136–7

behavioral problems (*cont.*)
  realistic expectations 162–3
  review of learning 130–2
  separation anxiety (SA) 137, 156–62
  systematic desensitization 130, 132–5, 139, 148, 150, 152, 156, 160–1
  unruly, rude, nuisance behaviors 137
Bekoff, Marc 39
Bird, Diana 83
bite inhibition, teaching 85
biting 68, 85, 141, 143, 237
body language 9, 16, 19, 20, 51–2, 56
  body posture 63–4
  ears 57, 61
  eyes 56–8
  fur/hair 57, 63
  mouth 57, 59–60
  tail 57, 61–3
Bones Would Rain from the Sky (2002) 13
breed differences
  behavior and 53
  caloric requirements and 182, 184
  exercise requirements and 168–169
  life expectancy 199
  neutering risks and 205–9
  weight guidelines and 176
Breland, Marian and Keller 80
British Association of Homeopathic Veterinary Surgeons 219–20
British Small Animal Veterinary Association (BSAVA) 67, 215
British Veterinary Association 212
BSAVA *see* British Small Animal Veterinary Association
BSAVA Manual of Canine and Feline Behavioural Medicine (2012) 215
Buchan-Smith, Hannah 25
burial 240
by-product hypothesis 33

CAAB *see* Certified Applied Animal Behaviorist
CADES canine dementia scale 234
cancer 206–8
cancer detection 47
Canine and Feline Behavior Management Guidelines (2015) 215
Canine Life Stage Guidelines (2012) 215
Case, Linda P. 98, 189
castration *see* neutering
The Cautious Canine booklet 155
Certified Applied Animal Behaviorist (CAAB) 11, 68, 128
Chaser (Border collie) 44–5

classical conditioning 75–7
  practical applications 77–8
Clothier, Suzanne 13
Code of Good Labeling Practice (UK) 185
cognition 31–2, 54
  application of knowledge concerning 47–8
  attachment 32, 36
  attention/perspective 34–5
  communication 32–3, 38
  communicative gestures 33
  determining validity of human expression 41
  effect of neutering on 210
  familiarity 37–8
  guilt 45–7
  human-dog synchronization 41–2
  importance of humans to dogs 32
  jealousy 42–3
  object-choice paradigm 33–4
  reading emotions 40–1
  social competence 3, 32–3, 38
  social referencing 36–7
  understanding human intentions 38–40
  word learning 43–5
cognitive dysfunction syndrome (CDS) 229, 231, 234
  causes 232
  changes in activity, anxiety, learning 233
  management/treatment 231, 233, 234
  symptoms 232
communication 32–3, 38
  facial expression/body posture 55
  interpreting emotional states to keep safe 64–9
  reading/responding to 56–64
  research into 53–4
  science of 52–6
counterconditioning 130, 134–6, 139, 148, 150–3, 155–6
crates 87, 88, 121, 159–60
cremation 240
The Culture Clash (2005) 27

DACVB *see* Diplomate of the American College of Veterinary Behaviorists
Darwin, Charles 51, 53
Dawkins, Marian Stamp 18
desensitization, systematic 132–4
diets 185
  grain-free/low carbohydrate 186–7
  raw food 185–6
  see also nutrition
Diplomate of the American College of Veterinary Behaviorists (DACVB) 11
Dodman, Nicholas 168
dog appeasing pheromone (DAP) *see* Adaptil

Dog Food Logic: Making smart decisions for your dog in an age of too many choices (2014) 189
Dog Sense (2011) 27
dogs -
   benefits of sharing life with 3–5
   domestication of 2–3, 33, 40, 94
   emotions of 9
   getting a new dog after loss 247
   human-dog bond 5–6
   information concerning 7–8
   making decisions/doing the right thing by 8–9
   numbers of 2
   understanding 9–11
DOGTV 121–2
domestication, theories of 2–3, 33, 40, 94
dominance myth 93–8
Donaldson, Jean 27
Dunbar, Ian 73, 141

ECAWBM *see* European College of Animal Welfare and Behavioural Medicine
elimination, inappropriate 46, 136, 157 (see also housetraining)
emotional state 64 (*see also* body language)
   aggressive 57, 58, 60, 61, 63, 64, 67–8, 107–8
   alert 66
   anxious/fearful 57, 58, 59–60, 61, 62, 63–4, 66–7
   aroused 66
   body language 56–64
   calm/relaxed 65
   excited/playful 65–6
   fearfully aggressive 68–9
   happy 56, 57, 59, 61, 62, 63, 64–5
   signalled by body part 56–64
emotions 9, 10, 19, 32, 51–2
   basic 39
   determining validity of expression 41
   guilt 45–7
   jealousy 42–3
   quality of life, influence on 105–6, 110, 120
   reading human emotions 40–1
   secondary 39 *see also* guilt, jealousy
   synchrony 41–2
   training, influence on 93
end-of-life decisions 225–7
   burial/cremation options 240
   decline/cognitive dysfunction 229, 231–4
   euthanasia decision/option 234–9
   euthanasia for behavior 236–238
   grief/coping with loss 240–6
   knowing when it is time 227–8, 230
   quality of life determination 228

enrichment
   access to kennels/crates 121
   animate 116
   behavioral benefits 123
   classical music 121
   definition 109–10
   importance of play 112
   inanimate 116, 118
   pheromones 122
   physical exercise 111, 113
   smell 122
   social and playtime 111
   Sodoku effect 110
   television for dogs 121–2
   toys, games, devices 114, 116, 118–21
   training 113–14
European College of Animal Welfare and Behavioural Medicine (ECAWBM) 128
European Pet Food Industry Federation 185
euthanasia
   for behavior issues 236–8
   decisions/options 234–6
   environmental factors 238
   in-home resources 241
   rehoming possibilities 238
   rule out medical/physical causes for behavior 237
   seek professional help 238
   suffering 238
   what to expect 239
exercise
   benefits 169–70
   need for 168
   reasons given for not walking dogs 171
   requirements 168–9
The Expression of the Emotions in Man and Animals (1872) 51, 53
exposure therapy *see* flooding therapy
extinction learning process 132

Farm Animal Welfare Council 17
fast mapping 44
fear 149–51
   as component of aggression 155–6
   medication 154
   recognizing/managing 151–2
   reducing 152–5
   socialization, importance of, in minimizing 149
Feisty Fido booklet 148
Five Freedoms 17, 18, 108–9, 227
   moving beyond 24–5
   providing 17–19
   and quality of life determinations 228
   test 21–4
   and training 90–3

flooding therapy, drawbacks of 135–6
flyball 114
food *see* diets; nutrition
foundation behaviors 147
Friedman, Erika 4

The Genius of Dogs (2013) 210
Grandin, Temple 19, 25
grief/coping with loss 240–2
   children and adolescents 242–4
   getting a new dog 247
   healing and moving on 244–5
   the Rainbow Bridge 245
   recommended reading 241, 242
   support on-line resources/hotlines 246
growling 58, 59, 60, 66, 67, 89, 94, 133, 143, 151, 238
Guidelines for Standards of Care in Animal Shelters (2010) 108–10
guilt 32, 45–7

habituation 72–3, 130
   practical applications 73–5
HAI *see* human-animal interaction
Hare, Brian 39, 210
health *see* mental health; physical wellness
Helsinki Chronic Pain Index 137
Horowitz, Alexandra 45
housetraining 46, 90
housing
   group/pair housing 19
   kennel/shelter accommodation 18, 118
"How to be the leader of the pack" (2007) 148
human-animal interaction (HAI) 4
human-dog relationship, models of
   dominance-submission (lupomorph) 36
   friendship 36
   parent-infant 36
   synchronization 41–2
Humane Society of the United States (HSUS) 212, 214

jumping up
   how to discourage 83–4, 138–9
   as a "nuisance behavior" 137–9
Karen Pryor Academy (KPA) 101
KPA *see* Karen Pryor Academy

learning *see* training
Levinson, Boris 4
life stages
   adolescent 199
   adult 199
   adult dogs during maintenance 183
   geriatric 199
   growth/reproduction 183
   puppy 199
   senior 199
low stress handling
   importance of 214–7, 255
   in veterinary care 196, 198, 214–7, 255–6

McConnell, Patricia 90, 148
Marder, Amy 15
Martin, Ken and Debbie 73
Mellor, David 24
mental health 105–6, 162–3
   daily routine 110
   enrichment 109–14, 116, 118–23
   environment 109
   factors influencing behavioral wellness 108–9
   responses to chronic stress 107–8
Miklósi, Adam 3, 250
Mills, Daniel 20
Muzzle Up! Project 144
muzzles 143–4

Neilson, Jacqui 133, 135, 143
neutering
   and cancer risks/benefits 206–8
   cognitive function and impairment with age 210
   and joint disorders 208–9
   and obesity 188, 209
   pediatric 210
   pros and cons 205–6
   and pyometra 209
   realities of living with intact dog 211
   summary 212
   surgical risk 210–11
   and urinary incontinence 209–10
nutrients 179
   carbohydrates 180
   fats 179–80
   minerals 181
   protein 179
   vitamins 180
   water 179
nutrition
   allergies 187
   canine athlete 183
   energy requirements 181–2
   feeding for life stage/energy level 182–3
   free-choice feeding 187–8
   how much to feed 188
   nutrients 179–81
   portion-control feeding 187
   pregnancy/lactation 184–5
   puppies/large breed puppies 183–4

requirements 179–82
Salmonella and *Escherichia coli* 186
senior dogs 185
setting feeding schedule 187–8
time feeding 188
weekend warrior 183
see also diets

obesity
   consequences 188–90
   and neutering 188, 209
   owner perception 190
   websites 191
object-choice paradigm 33–4
operant conditioning 75, 78–80
   behaviorist's perspective 80–2
   four quadrants 79–80, 83, 84–6, 131–2
origins of the domestic dog *see* dogs, domestication of
oxytocin 5

Panksepp, Jaak 10
Pavlov, Ivan 76, 77
Pet Food Manufacturers Association (PFMA) 176
The Pet-Friendly Practice: A guide for practitioners (2014) 215
Pet-Oriented Child Psychotherapy (1997) 4
PETS (Pets Evacuation and Transportation Standards) Act (2006) 26
PFMA *see* Pet Food Manufacturers Association
phobias 149
physical wellness 167–8
   activities/sports 171–6
   behavioral signs of underlying issues 136
   consequences of obesity 188–90
   diets 185–7
   exercise needs/benefits 168–71
   feeding guidelines 187–8
   feeding for life stage/energy level 182–5
   maintaining 190
   nutritional requirements 179–81
   pain 136–7
   weight guidelines 176–9
   weight management 191
play 63, 65–6, 75, 85, 111–2, 228
pregnancy/lactation 184–5
problems *see* behavioral problems
Pryor, Karen 81
punishment, positive, risks of 82, 87–93, 95, 97, 102, 127, 138, 145, 151, 161
punishment, positive vs. negative 131–2
puppies
   bite inhibition, teaching 85
   crate training of 88, 159

feeding and nutrition 182–4, 186
housetraining 90
influence of maternal stress on 149–50, 159
"nuisance behaviors" in 85, 137–8
socialization of 73–6, 149, 159, 203–4
training polite behaviors in 85, 138–40
vaccinations 200–4
veterinary care of 200, 210
Puppy Class 73–5
   enrolling in 75

QOL *see* quality of life
quality of life (QOL) 9, 225
   behavioral issues 162–3
   companion vs street dog life 14, 21–2
   dangers of anthropomorphism 13–14
   determining 228–9
   evaluating 17
   Five Freedoms 17–19, 21–5
   legalities 26
   as lifelong goal 16–17
   measuring/assessing 15–16
   minimizing negative influences 25
   providing 15
   recognizing/dealing with stress 19–21
   setting realistic expectations 26–8

reinforcement, positive vs. negative 131–2
reinforcement schedules 86
Reisner, Ilona 140, 145
response substitution 139, 147–8, 156
Rico (Border collie) 43–5
Royal Society for the Prevention of Cruelty to Animals (RSPCA) 128
RSPCA *see* Royal Society for the Prevention of Cruelty to Animals
Russian silver fox, tameness breeding experiment 3

SA *see* separation anxiety
safety recommendations for treating aggression
   avoid physical reprimands (positive punishment) 145
   avoid situations/triggers that provoke 145
   no interaction with animal 144–5
   seek help sooner rather than later 143
   warnings should be heeded 144
scents
   ability as human benefit 47
   detecting 47
   processing 47
Science magazine 2
Sdao, Kathy 97
secure base effect 6

senior dogs 185
separation anxiety (SA) 6, 137, 156–7
   book resources 162
   confirming 157–8
   owner role in 159–60
   symptoms of 156–7
   treating 160–2
"sit to say please", usefulness of 83–4, 140, 145
Skinner, B.F. 79, 80–1
social competence 3, 32–3, 38
social life of dogs 18–19
social referencing 32, 36–7, 37
socialization, importance of 73–6, 127, 140–1, 149, 203–4
socializing
   book resources 73
   definition 75
   goals 74
   timeline/sensitive periods 76
   tips 73
spaying *see* neutering
sports/activities 171
   agility 171–2
   disc dog 172
   dock jumping 172
   field trials 172
   flyball 173, 174
   lure coursing 173, 174
   musical canine freestyle/freestyle 175
   nosework 175
   sheepdog trial, herding event, dog trial 175
   sled dog racing 175
Stillwell, Victoria 83
stranger interaction 22, 37–8
stray/street dogs 14, 21–2
stress
   audit 23
   and children 22
   chronic 19–20
   dog parks 22
   effect on learning 20
   fenced yards 22
   kennel/shelter accommodation 18
   lowering 116
   and resilience 74
   shorter lifespan 19
   signs 16
   strangers 22
   tolerance for/responses to 20–1
synchrony 41–2
systematic desensitization 130, 132–5, 139, 148, 150, 152, 156, 160–1

terminology 4n–5n
toys/dog-friendly devices
   effectiveness 118
   Foobler 118
   food puzzle toys 120–1
   Kongs 118
   novel 118–19
   OmegaPaw "Tricky Treat" ball 118
   online sites 118
   Pet Tutor 118
   variety of 119–21
trainer, tips on selecting 101–2
training 27, 71–2
   agility 114
   appropriate 46–7
   associative learning 75–7
   bite inhibition 85
   body harnesses 92
   books, DVDs, websites 84
   canine scent work 114
   capturing 86
   changing problem behavior 138–40
   classical conditioning 77–8
   clicker training 81–2
   common mistake 97
   crate-training 87, 88
   dominance myth 93–8
   as enrichment 113–14
   enrolling in Puppy Class 75
   extinction 85
   fear and reliability 89–90
   flyball 114
   force-based 88, 92–3
   foundation behaviors 147
   habituation 73–5
   hanging/helicoptering 88
   head halters 92
   long-line/training leads 92
   lure-reward training 82
   management 86
   martingale collars 92
   no reward marker 91
   nothing in life is free/learn to earn programs 140
   operant conditioning 78–82
   Organic Training (K. Sdao essay) 97
   negative punishment 79–80, 83–4, 131, 139, 145
   negative reinforcement 79–80, 131
   positive punishment 79–80, 87–8, 89–90, 131
   positive reinforcement 79–80, 83–6, 131
   recall cues 97
   reluctance to relinquish older methods 92–3
   rewards 79, 81–2, 86
   rewards-based/force-free methods 90–2

scientific principles of dog learning 72–3
self-control 140
shaping 85
socializing 73, 74, 76
timing, importance of 81, 86
what really works 82–3
training tips
  be consistent 99
  be proactive not reactive 98–9
  choosing a trainer 102
  follow personal best instincts 101–2
  get professional help 101
  remember rules of reinforcement 99
  set realistic expectations 98
  set your dog up for success 100
  spontaneously reinforce preferred behavior 100
  use management wisely 101
trust 36–7

urinary incontinence 209–10 *see also* elimination, inappropriate

vaccinations
  core 201, 203
  importance of 199, 201
  killed vaccine 204
  modified live vaccine 204
  non-core 203
  recombinant vaccines 204
  schedules 202
  timing 203–4
veterinarian
  choosing 212
  fear of 217–19
  low-stress handling 214–17
  relationship with 197, 212, 214–15
veterinary care 195–6
  alternative medicine 219–20
  choosing a veterinarian 212, 214–19
  costs 196
  first 16 weeks 203–4
  importance of vaccinations 199, 201–3
  neutering 205–12
  protecting from disease 199, 201, 203
  recommended basic care schedule 196–9, 200
  reproductive health 205–6
  stress responses to 198
  transport to/from clinic 198

Waal, Frans de 10
Way to Go! How to housetrain a dog of any age 90
weight
  body condition score (BCS) 176, 177
  guidelines 176–7, 179
  muscle condition score 178
Weight Management Guidelines for Dogs and Cats (2014) 168
Westgarth, Carri 94–6
When Your Pet Dies: a guide to mourning, remembering and healing (2004) 241, 244–5
whining 59, 60, 85, 140, 149, 151, 156, 158
Wolfelt, Alan D. 244

yawning 59–60
Yin, Sophia 73, 215, 216